Samira **HAMZA REGUIG**
Hadil **YOUBI**

Presence and use of heterocycles for therapeutic purposes

Samira HAMZA REGUIG
Hadil YOUBI

Presence and use of heterocycles for therapeutic purposes

ScienciaScripts

Imprint

Any brand names and product names mentioned in this book are subject to trademark, brand or patent protection and are trademarks or registered trademarks of their respective holders. The use of brand names, product names, common names, trade names, product descriptions etc. even without a particular marking in this work is in no way to be construed to mean that such names may be regarded as unrestricted in respect of trademark and brand protection legislation and could thus be used by anyone.

Cover image: www.ingimage.com

This book is a translation from the original published under ISBN 978-620-6-71316-6.

Publisher:
Sciencia Scripts
is a trademark of
Dodo Books Indian Ocean Ltd. and OmniScriptum S.R.L publishing group

120 High Road, East Finchley, London, N2 9ED, United Kingdom
Str. Armeneasca 28/1, office 1, Chisinau MD-2012, Republic of Moldova, Europe
Printed at: see last page
ISBN: 978-620-7-65760-5

Presence and use of heterocycles for therapeutic purposes

Written by Dr. Samira HAMZA REGUIG

Mr^{elle} Hadil YOUBI

Table of contents

3

INTRODUCTION GENERAL

Approximately two-thirds of chemistry publications deal in some way with heterocycles, due to their structural diversity. Their properties and innumerable virtues. A very large number of substances and natural or synthetic drugs are in fact heterocycles. [1]

Heterocyclic chemistry is an integral part of organic chemistry. Heterocyclic compounds are widespread in nature and essential to life. The DNA of genetic material is also composed of heterocyclic bases - pyrimidines and purines.

Many heterocyclic compounds have agricultural applications as insecticides, fungicides, herbicides, pesticides, etc. They also have applications as sensitizers, developers, antioxidants, copolymers, etc. They are used as vehicles in the synthesis of other organic compounds. [2]

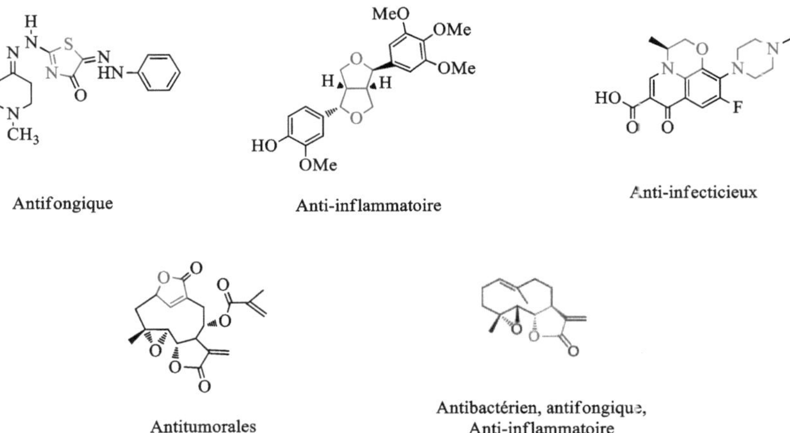

Figure 1: Examples of some heterocycles with biological activities.

In addition to an introduction and a general conclusion, this work is divided into three chapters, the first of which discusses in detail the systematic naming rules for heterocyclic systems in which several aromatic or heteroaromatic rings are fused, which is outside the scope of this thesis.

In the second chapter, we study the physical and chemical properties of each family according to the number of links, to define their influence on biological activity.

The therapeutic use of heterocycles is listed in the third chapter. It describes the multiple biological activities of each type of heterocycles of different heteroatoms.

Finally, a general conclusion will bring this work to a close, summarizing its most essential points.

CHAPTER I

HETEROCYCLE NOMENCLATURE

I-Introduction :

Heterocycles account for the majority of molecules used in industry, and are the subject of very active research worldwide. The literature devoted to heterocyclic chemistry is particularly extensive, with more than a quarter of current chemical publications relating to this field.

Their role in biological processes is of prime importance (vitamins, hormones, antibiotics, colorants, etc.) and they are also the basic structures of a wide variety of drugs.

II-Definition of heterocycles :

If the atoms form a chain, the corresponding compounds are said to be acyclic. On the other hand, if the atoms form a ring, the compounds are cyclic.

If the ring is made up entirely of carbon atoms, it's a carbocycle. A ring made up of two or more types of atoms is a heterocycle.

There are two groups of heterocycles: those containing one or more carbon atoms linked to one or more other elements such as oxygen, sulfur, nitrogen, etc., called heteroatoms, which are the organic heterocyclic compounds, and those containing no carbon atoms, the inorganic or mineral heterocycles, which are not included in our research. [[3] **(figure 1)**.

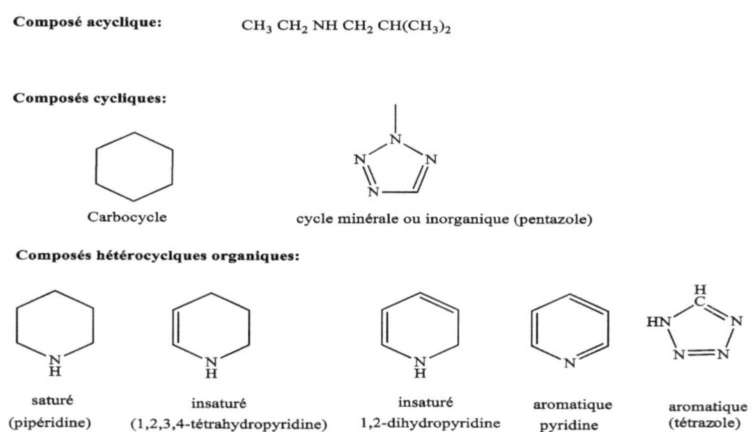

Figure 1: The different types of chemical compounds.

III-Nomenclature of heterocycles :

Heterocycle nomenclature is governed by international conventions defined by the International Union of Pure and Applied Chemistry **(IUPAC).**

Two main types of IUPAC nomenclature are used: **Hantzsch-Widman** and **alternative**. Hantzsch-Widman nomenclature rules apply to many compounds, particularly heterocycles with between three and ten ring atoms. For the rarer heterocycles **with** more than 10 ring atoms, a different nomenclature has been proposed. [3]

III.1-Hantzsch-Widman nomenclature rules:

In general, the chemical name of a heterocycle is constructed as follows:

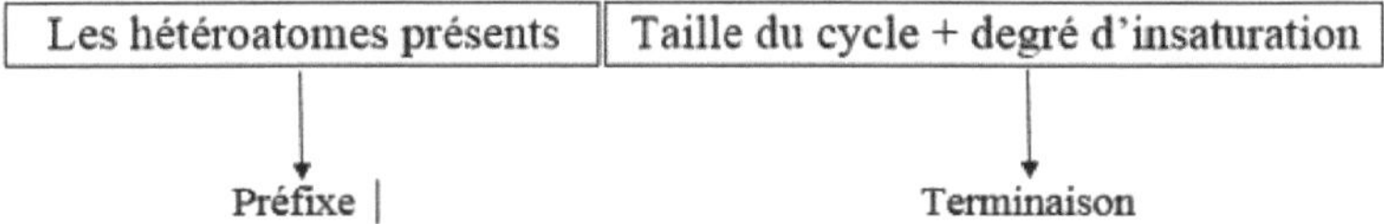

III.1.1-Rules for naming heterocycles: prefixes and suffixes

Each heteroatom is assigned a prefix. These prefixes are ordered according to a convention for naming a heterocycle. Table 1 shows the prefixes and their relative order (precedence of atoms $O > S > Se > N...$).

For example, a heterocycle with one nitrogen and one oxygen atom in its ring will have a name in which the prefixes are, successively, oxa (O), then aza (N) because $O > N$.

To make the name easier to read, we won't write "oxaaza" but oxaza, with the terminal "a" of the prefix elided before a vowel.

Hétéroatomes	Préfixes	Hétéroatomes	Préfixes
Oxygène (O)	oxa	Bismuth (Bi)	Bisma
Soufre (S)	thia	Silicium (Si)	sila
Sélénium (Se)	selena	Germanium (Ge)	germa
Azote (N)	aza	Etain (Sn)	stanna
Phosphore (P)	phospha	Plomb (Pb)	plomba
Arsenic (As)	arsa	Bore (B)	Bora
Antimoine (Sb)	stiba	Mercure (Hg)	mercura

Table 1

The number of ring members is indicated by two suffixes, one for unsaturated compounds, the other for saturated compounds.

Number of cycle links (cycle size)	Unsaturated cycle	Saturated cycle	
		non-nitrogenous	containing one or more N
3	irène	irane	iridine
4	ète	etane	etidine
5	ole	olane	olidine
6 (A series)	ine	donkey	
6 (B series)	ine	inane	
6 (C series)	inine	inane	
7	thorn	epane	
8	ocine	octane	
9	onine	onane	
10	ecine	ecane	

The suffix of the name of a fully unsaturated 6-membered ring containing several heteroatoms depends on which has the lowest rank in the order of precedence of the heteroatoms. Depending on whether this heteroatom belongs to one of the 3 series A, B or C, the suffix is **ine** or **inine**.

Serie A: O, S, Se, Te, Bi, Hg

Serie B: N, Si, Ge, Sn, Ph

Serie C: B, F, Cl, Br, I, P, As, Sb

Table 2

Example of nomenclature:

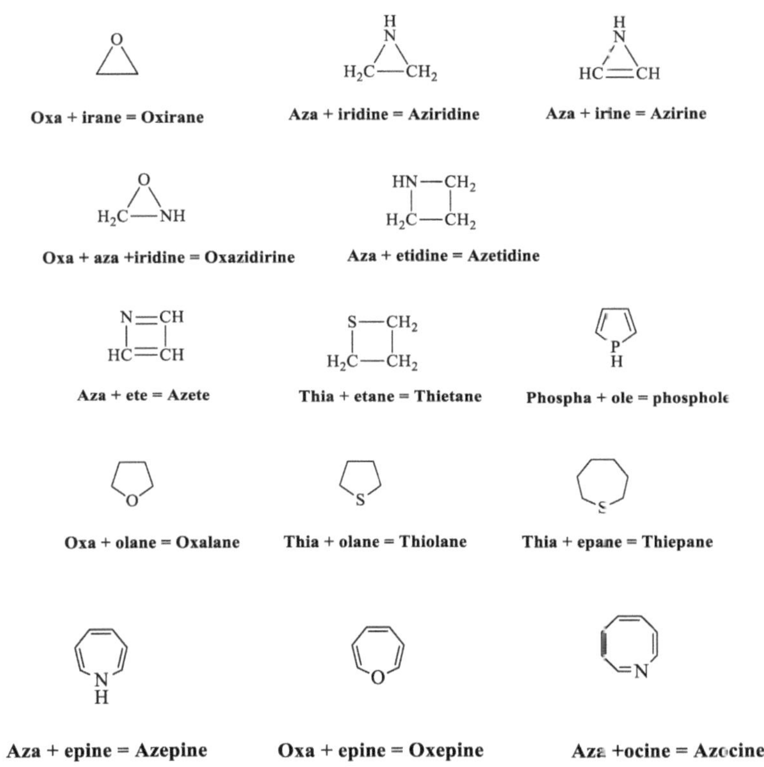

III.1.2-Monocycles with several heteroatoms of the same kind :

Monocycles containing several heteroatoms of the same kind are named by indicating the positions of each of them before the prefixes di-, tri-, tetra-...

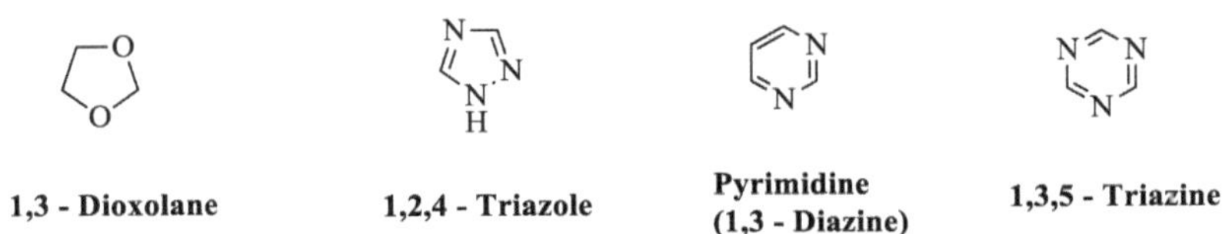

III.1.3-Monocycles with several heteroatoms of different kinds :

Monocycles containing several heteroatoms of different natures are named according to the precedence of the prefixes of each heteroatom and the number of each

13

(Table 1). Position 1 goes to the one with the highest precedence over the others (O > S > N..). [2]

Thia + aza + ole = Thiazole

(1,3 - Tiazole)

Oxa + aza + ine = Oxazine

(1,4 - Oxazine)

Thia + aza +ine = Thiazine

(1,4 - Thiazine)

III.1.4-Numbering :

1- With a heteroatom :

Numbering in the cycle starts from the heteroatom giving position 1 and the end closest to the branching (the branching index being the smallest possible index).

Pyridine

2,5 - Dimethylpyridine

3- Methyloxepin

Azocine

2- With two heteroatoms of the same nature :

When there are two or more identical heteroatoms in the ring, the numbers indicating the positions of these atoms are chosen so that their sum of indices is small for all heteroatoms and then for the substituents.

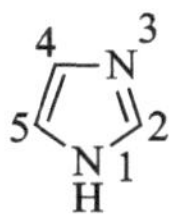

[1,3]

Numérotation correcte

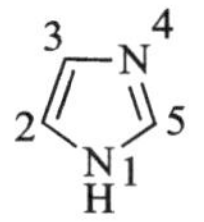

[1,4]

Numérotation incorrecte

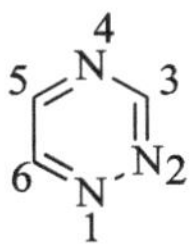

[1,2,4]

Numérotation correcte

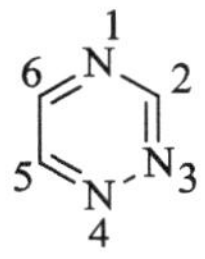

[1,3,4]

Numérotation incorrecte

3- With 2 different heteroatoms :

When the heterocyclic ring contains two or more different heteroatoms, the numbering starts from the heteroatom with the highest precedence over the others (O > S > N..).

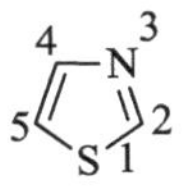

1,3 - Thiazole

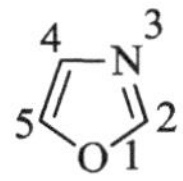

1,3- Oxazole

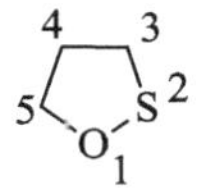

1,2 - Oxathiolane

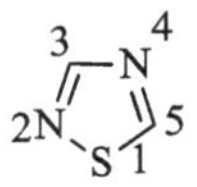

1,2,4- Thiadiazole (correcte)

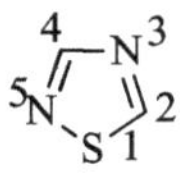

1,2,3- Thiadizole(incorrecte)

III.1.5-Partially saturated monocycles with a single heteroatom :

The following prefixes are used:

- Dihydro for hydrogenation of a double bond.

- Tetrahydro for hydrogenation of two double bonds.
- Hexahydro for hydrogenation of three double bonds.

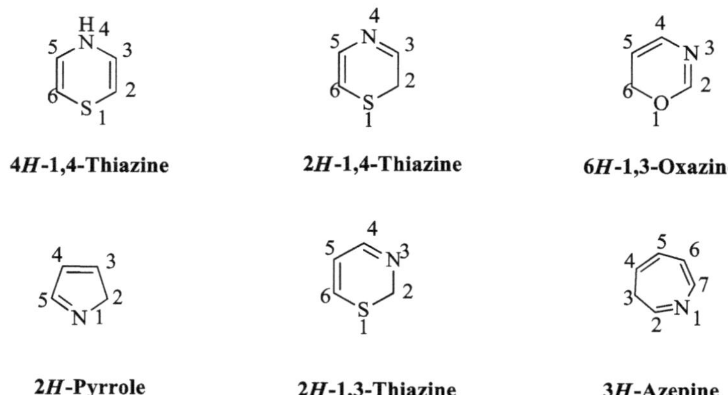

1,4-dioxine   **2,3-dihydro-1,4-dioxine**

III.1.6-Hydrogen positioning for certain structural isomers :

When several isomers differ in the position of a hydrogen in the ring, this is indicated by an italicized "*H*" preceded by the position of the atom to which it is bonded, this being the weakest if several possibilities exist [4].

4*H*-1,4-Thiazine   **2*H*-1,4-Thiazine**   **6*H*-1,3-Oxazine**

2*H*-Pyrrole   **2*H*-1,3-Thiazine**   **3*H*-Azepine**

III.2-Specific semi-systematic or semi-trivial nomenclature :

For many natural compounds discovered long before the publication of the I'lUPAC rules, specific nomenclatures are often still in use (trivial nomenclature). This is the case for certain structures (**pyrrole, pyridine, quinoline...**) and natural products, including a large number of alkaloids (morphine, cocaine...). Some nomenclatures are said to be semi-systematic or semi-trivial insofar as part of the name refers to a systematic suffix: pyrrolidine, morphinane, tropane, tropanol, tropinone. [5]

III.2.1-Some examples illustrating the nomenclature :

Selenophene

Tellurophene

1*H*-isomère

Pyrrole

Furan

Thiophene

Pyridine

Pyridazine

Pyrimidine

Pyrazine

Pyran

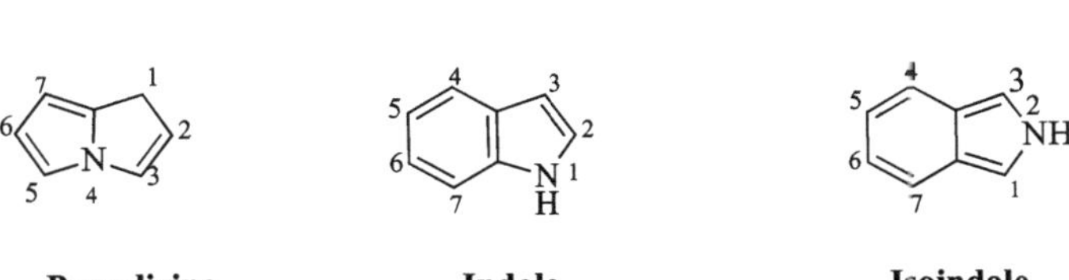

Pyrrolizine

Indole

Isoindole

Quinoline

Isoquinoline

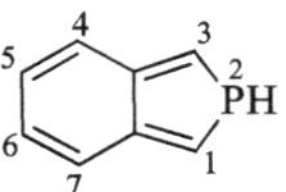

Phosphindole

Isophosohindole

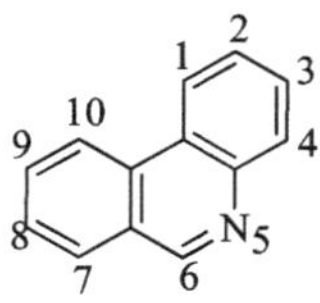

Phenanthridine

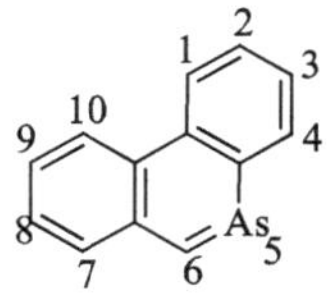

Arsanthridine

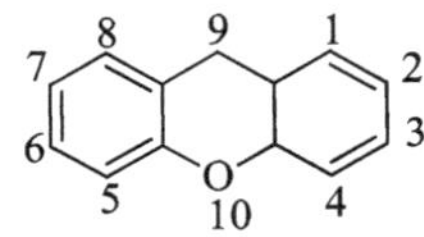

Xanthene

Exception de numérotation

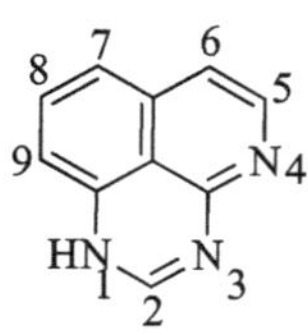

Perimidine

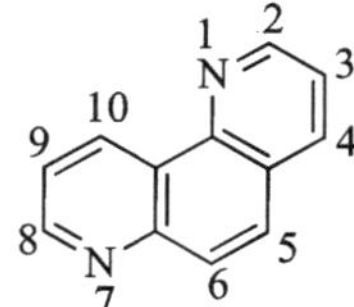

Phenanthroline

III.2.2-Bicyclic system: fused heterocycles

The nomenclature of fused heterocycles can be systematic or trivial. The system of fused heterocycles is constructed by the fusion of two or more structural rings, the ring must have the maximum number of non-cumulative doublets and fused in such a way that each ring has one bond in common with the other. [2]

III.2.2.1-Nomenclature rules in this system :

1- The fused heterocyclic system is made up of two types of cycle, one of which is the main cycle and the other(s) the secondary cycle(s).

Benzothiazole   **Benzene**   **Thiazole**

Benzimidazole   **Benzene**   **Imidazole**

2- The trivial nomenclature is the most favorable in the case of fused heterocycles, as indicated in the previous table, but if it doesn't exist in the latter the systematic nomenclature will be used.

3- The base component must be a heterocycle, and if we have a choice this base is determined by the order of preference.

III.2.2.2-How to select the main cycle ?

  a- **The nitrogenous heterocycle:** when there is a nitrogenous heterocycle in this system, it is preferable to choose it as a main cycle.

cycle principal : Pyridine   **cycle principal : Pyrrole**

  b- **A heterocycle other than the nitrogen heterocycle:** if a heterocycle is present, it is chosen as a main ring in all cases, according to the rules of this system.

If there are two heterocycles different from the nitrogen heterocycle, the main ring is the one containing the heteroatom according to the relative order of the atoms in the periodic table (precedence of O > S > Se > N atoms).

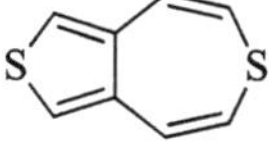

cycle principal : Furan

c- A heterocycle with a large number of rings, such as **quinoline**; it's the main ring if it's flanked by another heterocycle with fewer rings.

cycle principal : Quinoline

d- Two heterocycles of different sizes, the larger cycle takes priority.

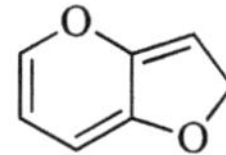

cycle principal : Thiepine   **cycle principal : Pyran**

e- Heterocycles with the same number of members but different numbers of heteroatoms: the heterocycle with the greatest number of heteroatoms is chosen as the main ring.

cycle principal :Oxazole

Heterocycles with the same number of members and the same number of different heteroatoms: in this case, the main ring is the one with heteroatoms according to atomic precedence (O > S > Se > N).

cycle principal :Oxazole   **cycle principal : Thiazole**

f- Heterocycles with the same number of members and the same heteroatoms: the base compound is chosen according to the end closest to the heteroatom position of the head.

cycle principal : Pyridazine   **cycle principal : Pyrazole**

The secondary cycle is considered a prefix, with the e ending replaced by **a** and the "a" elided before a vowel.

Pyrazine Pyrazino

Pyrazole Pyrazolo

Thiazole Thiazolo

But there are exceptions to this nomenclature system for some heterocycles presented in the following table:

the prefix heterocycle
Pyridine Pyrido-
Quinoline Quino-
Isoquinoline Isoquino-
Furan Furo-
Thiophene Thieno-

Imidazole Imidazo

Table 3: Prefixes of some common heterocycles

Main ring bonds are indicated by italicized alphabetical letters starting with *'a'* for bond 1,2, *'b'* for bond 2,3, *'c' for* bond 3,4 and *'d'* for bond 4,5 and so on.

Secondary-cycle atoms are numbered in the usual way; 1, 2, 3, 4, 5, *etc.,* according to the precedence of the atoms.

Atoms common to both cycles (fusion side) are indicated by the appropriate letters and numbers, and are enclosed in a square bracket immediately after the secondary component prefix [4].

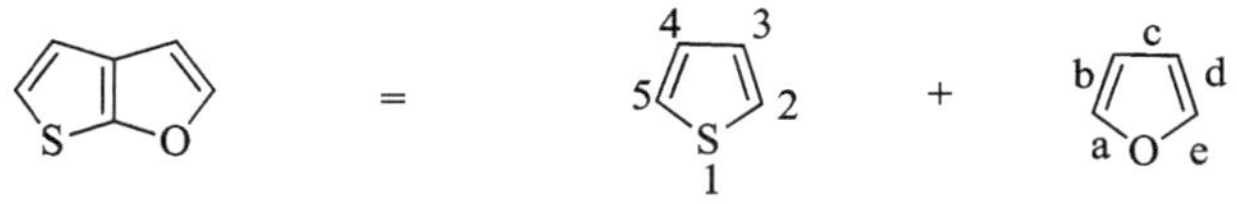

Thieno[2,3-*b*]furan = **Thiophene** cycle secondaire + **Furan** cycle principal

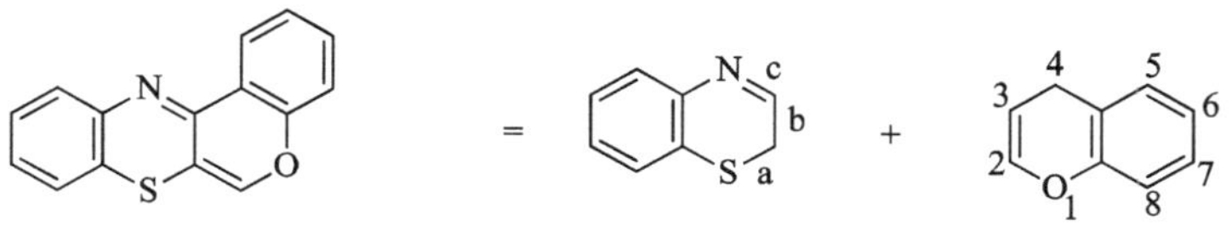

Benzopyrano [3,4-*b*] benzithiazine = **[1,4] Benzothiazine** cycle principal + **Benzopyran** cycle secondaire

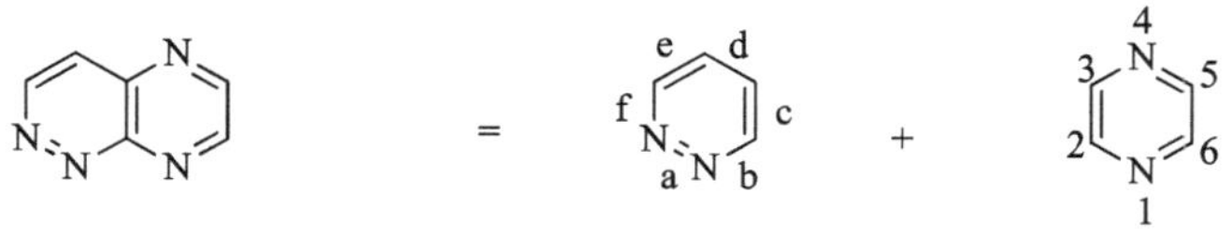

Pyrazinol[2,3-*c*]pyridazine = **Pyridazine** cycle principal + **Pyrazine** cycle secondaire

A common heteroatom: If a fusion position is occupied by a heteroatom, both components (ring systems) are considered to possess this heteroatom.

Imidazo[2,1-*b*]oxazole   **Imidazole**
cycle secondaire   **Oxazole**
cycle principal

III.2.2.3- Numbering

a) The fused heterocyclic system is numbered independently of the ring combination (primary and secondary). Numbering starts from the atom adjacent to the bridgehead position with the lowest possible substituent(s) on the heteroatom(s).

Benzo[*b*]furan   **3,1-Benzooxaepine**

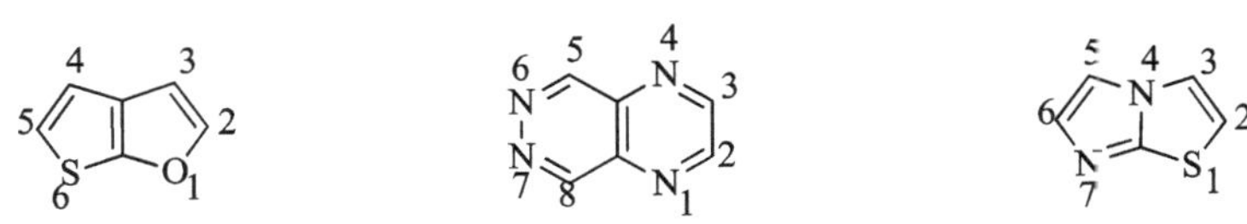

Thieno[2,3-*b*]furan   **Pyrazino[2,3-*d*]pyridazine**   **Imidazo[2,1-*b*]thiazole**

b) The carbon atom common to two rings is placed in the lowest possible position, but not numbered. However, the heteroatom at a position where two rings merge (common heteroatom) is numbered.

Imidazo[1,2-*b*]pyridazine   **1,2,4-Trizolo[4,3-a]pyridine**

c) The position of a saturated atom is indicated by *italicized* hydrogen and receives the smallest possible number of the function group position in the ring.

23

$2H$ - Furo[3,2-*b*]pyran

III.2.2.4- Heterocycle attached to a benzene ring :

The name of the heterocycle is preceded by the prefix **"benzo"** (with elision of the "o" before a vowel) followed by a bracketed letter designating the bond common to the two rings defined from the heterocycle.

3,1-Benzoxepine   **4*H*-1,4-Benzothiazine**   **4*H*-3,1-Benzooxazine**

III.3- Replacement parts list :

Heterocycles are considered to be derived from carbocyclic compounds by replacing one or more carbon atoms with the heteroatom(s). This forms the basis of replacement nomenclature for heterocycles and, consequently, the rules of replacement nomenclature are similar to those applied for carbocyclic compounds. This is the most systematic nomenclature and is used for heterocycles containing unusual heteroatoms. [6]

III.3.1-Monocyclic heterocycles :

a) As all prefixes end with the letter 'a', the replacement nomenclature is also called 'a' nomenclature. The position and prefix of each heteroatom are placed before the corresponding carbocyclic name.

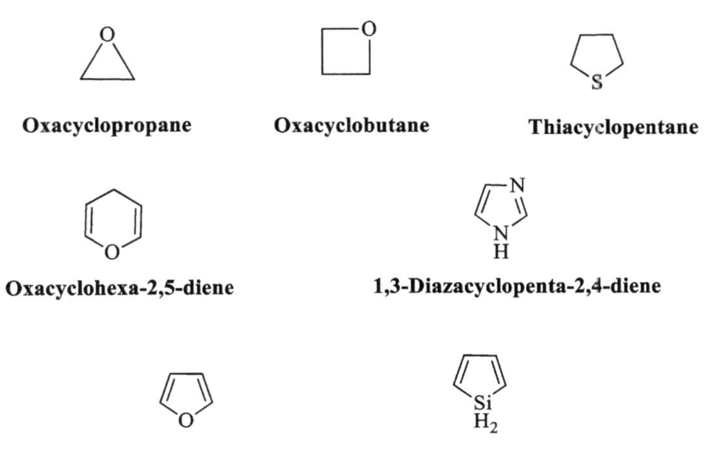

Oxacyclopropane   **Oxacyclobutane**   **Thiacyclopentane**

Oxacyclohexa-2,5-diene   **1,3-Diazacyclopenta-2,4-diene**

Oxacyclopenta-2,4-diene   **Silacyclopenta-2,4-diene**

b) The benzene-derived replacement nomenclature is used when there are double bonds in the ring.

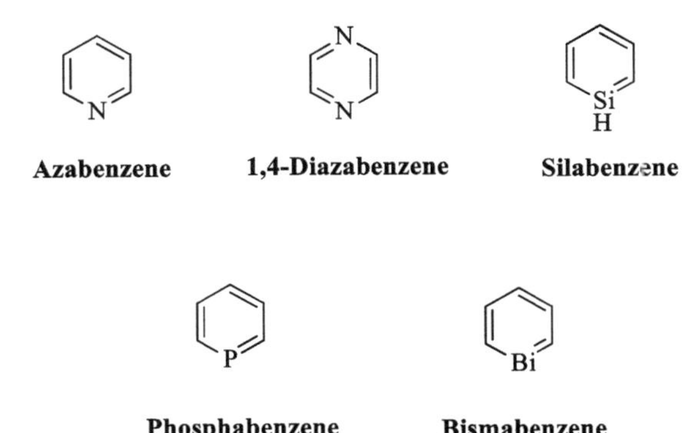

Azabenzene   **1,4-Diazabenzene**   **Silabenzene**

Phosphabenzene   **Bismabenzene**

Numbering according to :

1. The heteroatom with superior precedence over the others

2. Other heteroatoms according to table

3. Double bonds

4. Substituents

25

1-Oxa-3-azacyclopenta-2,4-diene

1-Thia-4-azacyclohexa-2,5-diene

1-Thia-4-aza-2silacyclohexane

1,4-Dithiacyclohexa-2,5-diene

III.3.2-Attached heterocycles :

The replacement nomenclature for these heterocycles is based on the following rules:

1. The positions and prefixes of the heteroatoms are written at the beginning of the text.

2,5-Diazaanthracene

3,9-Diazaphenanthrene

2. The junction heteroatom receives an individual number in the fusion nomenclature system, whereas in the replacement nomenclature system, the junction heteroatom receives the same label as an adjacent non-connective carbon atom, but with an 'a' or 'b' suffix.

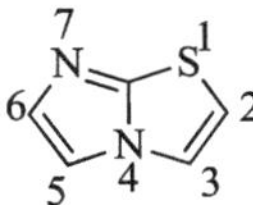

Imidazo[2,1-b]thiazole
nomenclature de fusion

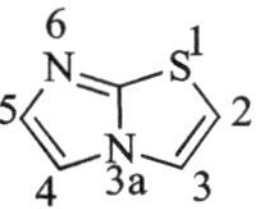

1-Thia-3a,6-diazapentalene
nomenclature de remplacement

3. The corresponding carbocyclic ring without the maximum number of non-cumulative double bonds is named without using a prefix for the hydrogen atom.

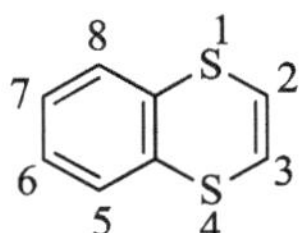

2-Oxa-3-thia-1,5-diazaindane

4. If the corresponding carbocyclic ring does not have the maximum number of non-cumulative double bonds, the heterocyclic system is considered to have the maximum number of conjugated or free double bonds. The corresponding carbocycle is named because it contains the maximum number of non-cumulative double bonds.

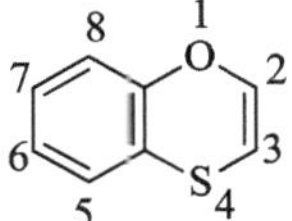

1,4-Dithianaphthalene

1-Oxa-4-thianaphthalene

III.3.3 - Spirane nomenclature :

1. Compounds in which two rings are fused at a common point are known as spiro compounds, and the common atom that is quaternary in nature is referred to as the spiro atom. Spiro compounds can be classified according to the number of spiro atoms; (1) monospiro- (2) dispiro- and (3) trispiro ring systems.

The nomenclature is based on the following rules:

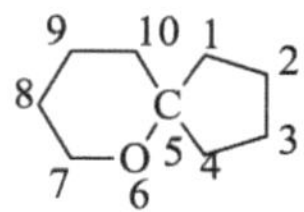

spiro hydrocarbon

Spiro[x,y]alkane

x = le nombre des atomes à part l'atome spiro
dans le plus petit cycle

y = le nombre des atomes à part l'atome spiro
dans le plus grand cycle

↓

spiro[4,5]décane

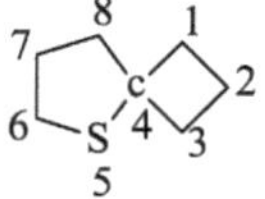

Spiro heterocycle

Spiro[x,y]alkane

x = 4 (le spiro atome n'est pas compté)
y = 5 (le spiro atome n'est pas compté)
alkane:le nombre de tous les atomes present (méme ci l'hétéroatome) = 10 atome ⟶ décane

le préfixe de l'hétéroatome: oxa

↓

6-Oxaspiro[4,5]décane

2. Numbering starts from the smallest ring atom (if the rings are of different sizes) attached to the spiro atom and proceeds first around the smaller ring, then around the larger ring through the spiro atom.

5-Thiaspiro[3.4]octane   **5-Oxa-9thiaspiro[3.5]nonane**

3. The heterocyclic ring is preferred to the carbocyclic ring of the same size. If both rings are heterocyclic, preference is given to the heterocycle with heteroatom appearing first in **Table 1.**

2-Oxaspiro[5.5]décane   **1-Oxa-6-thiaspiro[4.4]nonane**

4. If unsaturated heterocycles are present, the nibbling pattern remains the same, but the direction around the ring such that the multiple bond is given as low a number as possible.

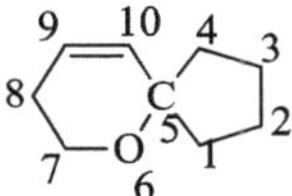

1-Oxaspiro[4.5]déc-6-ène   **6-Oxaspiro[4.5]déc-9-ène**

IV-Conclusion :

Many organic compounds, including heterocyclic compounds, have trivial names. This is usually due to the occurrence of the compound, its first preparation or its specific properties.

We conclude that the purpose of nomenclature is to provide us with unambiguous names for heterocycles, so that we can distinguish between them; each one has its own name. On the other hand, a heterocycle can be given several different names according to other nomenclature systems, but the system currently in use is trivial nomenclature, since it is the most simplified of the others.

CHAPTER II

PROPERTIES OF HETEROCYCLES

I-Introduction:

As well as the type of atoms in the ring, their total number is also important, as this determines the size of the ring. The smallest ring is the three-membered ring. The largest rings are the five- and six-membered heterocycles. There is no upper limit; there are heterocycles with seven, eight, nine and more.

To determine the stability and reactivity of heterocyclic compounds, it is useful to compare them with their carbocyclic analogues. In principle, each heterocycle can be derived from a carbocyclic compound by substituting a heteroatom at the appropriate CH_2 or CH group. [7]

II-Aromaticity of heterocycles :

The classification of unsaturated heterocycles depends on heteroatom participation in the pi electron system. Consequently, the reactivity of heterocycles is directly proportional to the number of heteroatoms. So that which contains a single heteroatom is the most stable. [8]

II.1- Heterocycle aromaticity conditions :

An organic compound is said to be aromatic when it satisfies the following conditions:

1- Presence of a ring with a conjugated π-system, formed by double bonds and/or non-bonding doublets;

2- Each atom in the ring has a p orbital;

3- The p orbitals overlap (conjugated π system), the molecule being planar at the level of this cyclic compound (sp2 hybridization).

4- The delocalization of π electrons leads to a decrease in the molecule's energy

If the first three criteria are met, but delocalization would lead to an increase in energy, the compound is said to be anti-aromatic.

In practice, the 4[e] criterion is translated by Hückel's rule: delocalization leads to a decrease in the molecule's energy (and therefore to its stabilization) if the number of electrons π is equal to $(4n + 2)$, where *n is* a positive integer or zero. [9]

II.2-Aromaticity of the benzene ring:

We're going to study the aromaticity of benzene to understand :

In the case of benzene, the carbon atoms are sp-hybridized[2] , and the hydrogen atoms are in the same plane as the carbon. This leaves us with 6 orbitals containing 6 pi-electrons. Consequently, benzene is an aromatic compound since it has 4n+2 pi electrons (with n =1).

There are two mesomeric forms for benzene, Y has a total of six p-orbital electrons that form the stabilizing electron clouds above and below the aromatic nucleus This diagram shows one of the molecular orbitals containing two of the delocalized electrons, the other molecular orbitals are almost never drawn.it can also be represented by the shape that represents the cyclic arrangement of electrons **(figure 2)**.

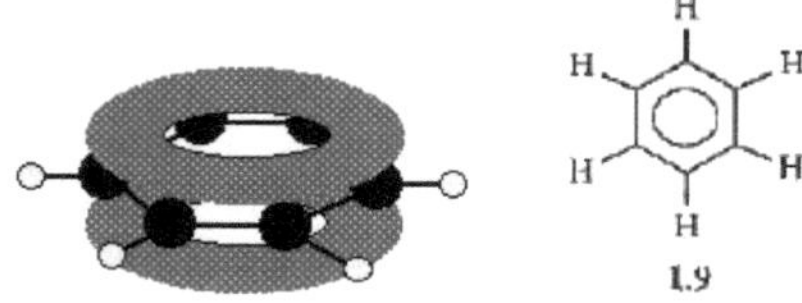

Figure 2: Aromaticity of the benzene ring

Due to the aromaticity of benzene, the resulting molecule has a planar shape, with each C-C bond having a length of 1.39 Å and a bond angle of 120°. But how is it possible to have all the bonds have the same length if the ring is conjugated both single (1.47 Å) and double (1.34 Å), so it's important to note that there are no distinct single or double bonds in benzene. On the contrary, delocalization of the ring means that each bond counts as one and a half bonds between the carbon atoms, which makes sense because, experimentally, we find that the actual bond length lies between a single and a double bond.

II.3-Why the (4n+2) π-electron rule?

According to Hückel's molecular orbital theory, a compound is particularly stable if all its molecular bonding orbitals are filled with paired electrons. This is the case with aromatic compounds, i.e. they are quite stable.

With aromatic compounds, 2 electrons fill the lowest-energy molecular orbital, and 4 electrons fill each subsequent energy level (the number of subsequent energy levels is indicated by n), leaving all bonding orbitals filled and no anti-bonding orbitals occupied. This gives a total of 4n+2 electrons.

Benzene has 6 electrons; its first 2 electrons fill the lowest energy orbital, and it has 4 electrons left. These 4 fill the orbitals of the next energy level. Notice how all its bonding orbitals are filled, but none of the anti-bonding orbitals have electrons **(Figure 3)**.

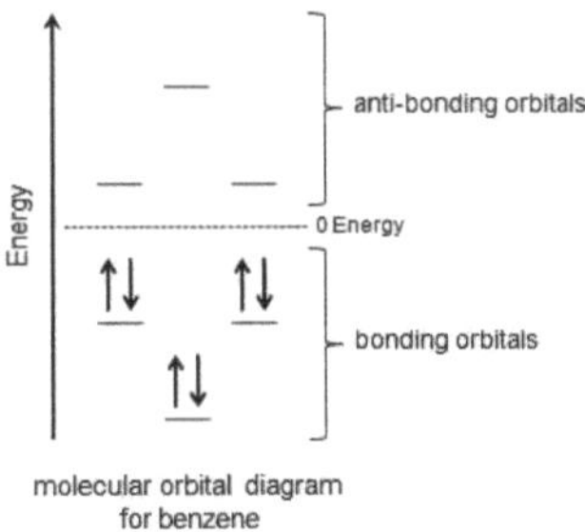

Figure 3: Molecular orbital diagram of the aromatic compound

To apply the 4n+2 rule, first count the number of π electrons in the molecule. Then, set this number to 4n+2 and solve for n. If n is 0 or any positive integer (1, 2, 3,...), the rule is respected. For example, benzene has six electrons:

4n+2=6

4n=4

n=1

For benzene, we find that *n=1*, which is a positive integer, so the rule is satisfied. [10]

III-Reactivity of heterocycles:

III.1-Three-membered heterocycles :

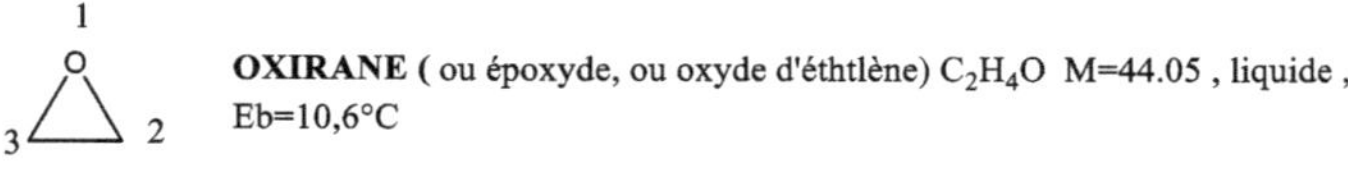

OXIRANE (ou époxyde, ou oxyde d'éthtlène) C_2H_4O M=44.05 , liquide , Eb=10,6°C

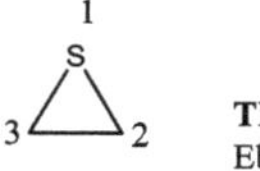

THIIRANE (ou épisulfure, sulfure d'éthylène) C2H2O M= 60.11 , liquide , Eb=55-60°C , avec décomposition

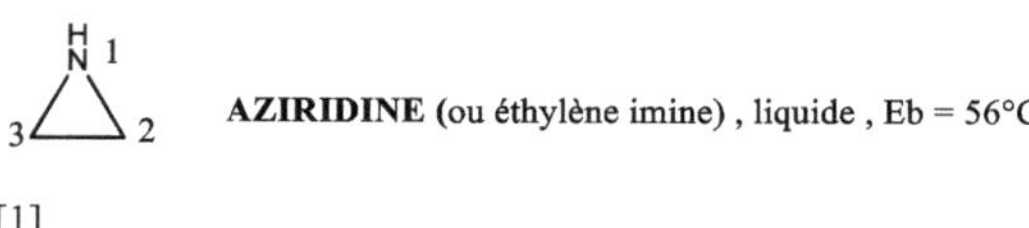

AZIRIDINE (ou éthylène imine) , liquide , Eb = 56°C

[1]

Three-membered heterocycles are tense molecules, with dihedral angles close to 60°, i.e. far from109°28 '. These heterocycles have high energy contents compared with their acyclic isomers, and generally have a short bond (C-C) compared with cyclopropanes (thiirane is taken as an exception).

These compounds are highly reactive with acids, bases and nucleophilic reagents such as amines, alcohols... This explains their high toxicity and carcinogenic potential.

In the case of unsaturated heterocycles, the introduction of the double bond further increases angular distortion, and this inevitably increases ring deformation.

The chemistry of three-membered heterocycles is dominated by the presence of ring constraint in these molecules. The ring constraint leads to enhanced reactivity and facilitates ring-opening reactions. The stability and overall reactivity of these heterocycles are not only attributed to the combined effect of bond shortening and angular distortion, but also depend on the presence of heteroatom. [2]

III.1.1-Thiirane

Thiirane is a three-membered saturated sulfur heterocycle, also known as thiacyclopropane, and was first synthesized by *Staudinger* and *Pfinniger* in 1916. This

ring system is present in various natural products in isolated form. Due to the larger atomic radius of the sulfur atoms and the smaller ring size, the geometry of the molecule is similar to an acute-angled triangle [3].

III.1.1.1-Physical properties

Thiirane (ethylene sulfide) is a colorless liquid with a boiling point of 55°C and low solubility in water. Even in the absence of light, it forms polymers as a result of cycle opening. This instability also explains the absence of this cycle among natural products. [2]

III.1.1.2-Chemical properties :

Due to their high ring voltage, these small heterocycles play an important role in the easy synthesis of organic products, pharmaceuticals and intermediates for the construction of various natural products because of their high reactivity.

Among three-membered heterocycles, aziridines are a particularly versatile class of heterocycles useful for building a variety of molecules with vivid biological activities. The highly compressed bond angles of three-membered heterocyclic compounds make them highly reactive to nucleophilic, electrophilic, thermal and photochemical reactions.

III.1.1.2.1-Cycle opening reactions :

III.1.1.2.1.1-By nucleophilic reagents :

Ring opening by nucleophiles is a stereospecific reaction involving inversion of the configuration of the least-substituted carbon (by electron-donating groups). The nucleophile's attack is directed away from the sulfur atom, leading to a sulfide ion or thiol, depending on the nucleophile **(Scheme 1)**.

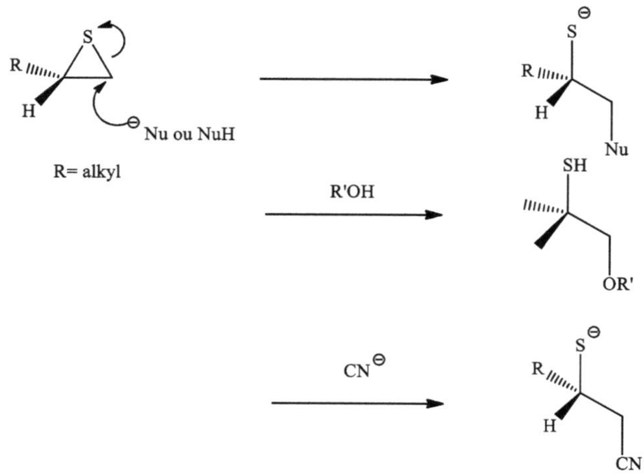

Diagram 1

III.1.1.2.1.2-By lithium aluminum hydride :

Lithium aluminum hydride opens the ring, leading to a thiol. The reaction is stereospecific, as shown by the reaction carried out with lithium aluminum deuterate (SN_2) **(Scheme 2)**.

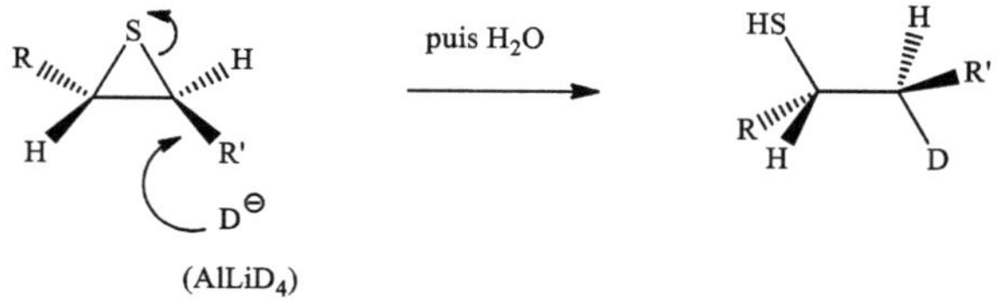

Diagram 2

III.1.1.2.2.3-Hydric acid halos :

Concentrated hydrohalic acids protonate the sulfur, then the halide ion reacts on the least-substituted carbon to give p-halogenothiols **(Scheme 3)**.

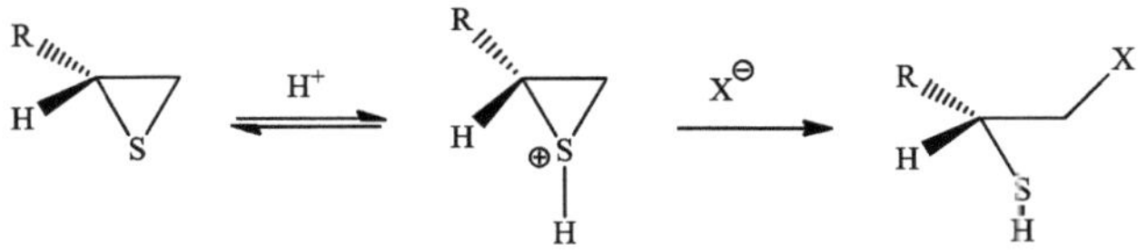

Diagram 3

III.1.1.3-Oxidation

There are two oxidized forms of thiirane, monoxide and dioxide. Thiiranes are oxidized to thiirane monoxides by peroxyacids or sodium periodate. These oxides decompose on heating to give alkenes and sulfur monoxide **(Scheme 4)**.

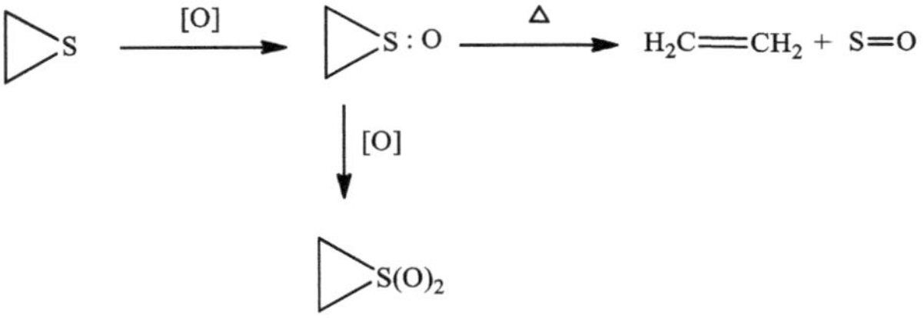

Diagram 4

III.1.1.4.1 -Desulfurization with n-butyllithium :

Addition of n-butyllithium to 1,2-dimethylthiirane at -78°C in THF leads to but-2-ene **(Scheme 5)**.

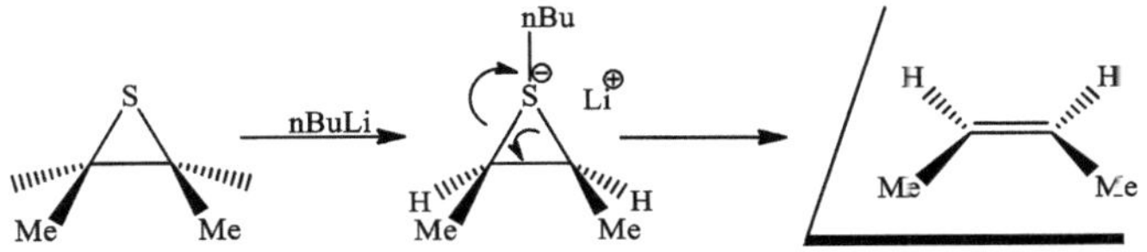

Diagram 5

III.1.1.4.2 -Desulfurization with triphenylphosphine :

Thiiranes substituted with a ring-conjugated aromatic or ethylenic group are thermally desulfurized.

Triphenylphosphine desulfurizes thiiranes (A). This mechanism begins with its coordination to the thiirane sulfur, followed by ring opening, and the formation of the ethylenic (B) from which the heterocycle is derived. [3] (**Scheme 6**)

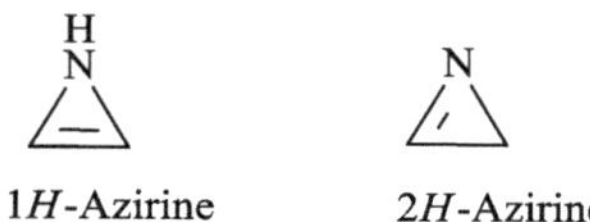

(A)   (B)

Diagram 6

III.1.2- Azirines :

1*H*-Azirine   2*H*-Azirine

Azirine, also known as azacyclopropene, is a three-membered unsaturated nitrogen heterocycle. Being highly reactive molecules, azirines are involved in a variety of chemical transformations, acting as dienophiles, dipolarophiles, electrophiles and nucleophiles.

Azirines are classified into *1H-azirines* with a C=C bond and 2H-azirines with a C=N bond. *1H-azirine* has short-lived stability. However, *2H-azirine* is comparatively more stable.

III.1.2.1-Physical properties :

Azirines are poultry-smelling liquids that are irritating to the skin. The C=N and C-N bond lengths in azirine are 1.270 and 1.527 Å, respectively, while the C-C bond length is 1.458 Å, reduced from the normal C-C bond length.

III.1.2.2-Chemical properties :

2H-azirines have been found to undergo a wide variety of transformation reactions:

1. Nucleophilic reactions

2. Electrophilic reactions

3. Thermal cleavage

4. Photochemical reactions

5. Diels-Alder reactions

Let us note that the nucleophilic reactions in our study :

III.1.2.2.1 -Nucleophilic reactions :

2H-azirines are highly sensitive to nucleophiles and the presence of electron-withdrawing groups on the carbon atom also activates the ring for nucleophilic addition reactions followed by ring opening.

In the reaction of 2H-azirine with benzylamine, the initial step is the addition of amine to the C=N bond, followed by ring cleavage to open the ring.

$$Ar\text{-azirine-COOMe} \;+\; C_6H_5CH_2NH_2 \;(\text{Benzylamine}) \longrightarrow MeOOC\text{-}C(NHCH_2C_6H_5)=C(Ar)\text{-}NH_2$$

III.2-4-membered heterocycles :

OXETANE — liquide , Eb =47,6°C — C_3H_6O M=58,08

THIETANE — liquide, Eb= 95°C — C_3H_6S M=74.14

AZETIDINE — liquide, Eb=63°C — C_3H_7N M=57.09

OXETAN-2-ONE — liquide , Eb=162°C — $C_3H_4O_2$ M=72,06

AZETE — Solide , Eb=37°C — C_3H_3N M=53

The reactions of these heterocycles are similar to those of the corresponding 3-membered heterocycles, but are made more difficult by the reduced ring tension. Most of these reactions lead to ring opening.

Quite generally, in acidic environments, the SN_1 mechanism is the most frequent. The results of these reactions are variable and depend very much on the nature and number of substituents on the ring.

III.2.1 -Oxetane :

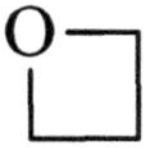

Replacing one of the CH_2 groups in cyclobutane rings with an oxygen atom produces an oxetane ring, also known as oxacyclobutane. The oxetane structure is slightly distorted and the bond angle to the O atom is 92 degrees. [5]

III.2.1.1- Physical properties

Oxetane is a colorless liquid soluble in water and most organic solvents; it has a boiling point of 47-48°C.

III.2.1.2 -Chemical properties :

III.2.1.2.1 -Electrophilic substitution reactions :

Oxetanes undergo electrophilic substitution with different electrophiles under different reaction conditions, yielding numerous products to generate molecular diversity. The different reactions are illustrated in the following diagram (**Figure 7**).

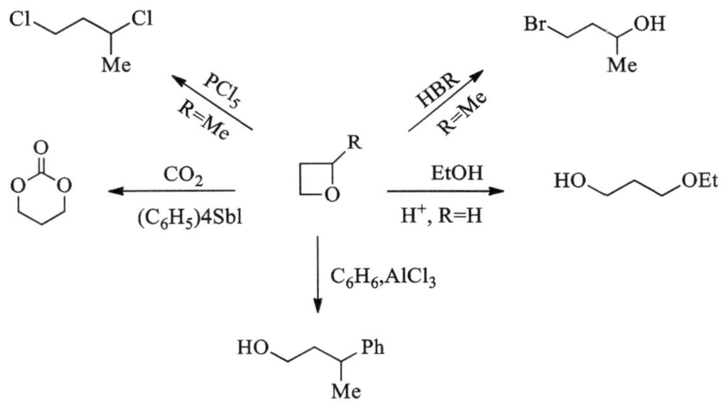

Diagram 7

III.2.1.2.2 -Nucleophilic substitution reactions :

The oxetane ring is prone to nucleophilic substitution reactions with different nucleophiles and produces ring-opened products as illustrated in the following diagram. [5] (**Scheme 8**)

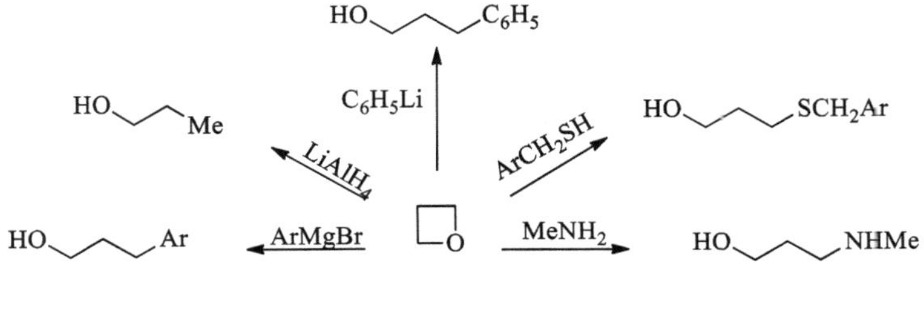

Diagram 8

$$\underset{\text{(1)-Azetine}}{\boxed{}=N} \qquad \underset{\text{(2)-Azetine}}{\boxed{}-NH}$$

Azetine is a four-membered unsaturated heterocycle with three carbons, one nitrogen and one double bond in a ring system. Azetine exists in two isomeric forms: (1) 1-azetine and (2) 2-azetine, which differ only in the position of the double bond. In the 1-azetine ring, the nitrogen is in the imine form (N=C), while in 2-azetine it is in the secondary amine form (NH). [4]

III.2.2.1-Physical properties :

Azetine is highly unstable and a gas at room temperature. It can be trapped at the temperature of liquid nitrogen. It is a colorless liquid at -70°C and polymerizes rapidly even in a sealed tube degassed at 20°C. The half-life of azetine in a sealed tube dissolved in $CFCl_3$ is around 90 min, or 3 days in the presence of hydroquinone. The presence of traces of oxygen and acid induces rapid polymerization.

III.2.2.2-Chemical properties :

Reduction of azetine with $LiAlH_4$ in ether at 0°C gave azetidine. However, azetine in the presence of HCN in methylene chloride at -50°C gave 2-cyanoazetidine as adduct. Pyrolysis of 1-azetine produces 2-aza-1,3-butadiene. (**Scheme 11**)

Diagram 11

Disubstituted 2-ethoxy-3,3-azetine is obtained by alkylation of 3,3-disubstituted azetidin-2-one with triethyloxonium tetrafluoroborate salt, treatment with concentrated mineral acid gave 3,3-disubstituted azetidin-2-one, in the presence of aqueous diethyl acid, disubstituted malonate was isolated. Reduction with LiAlH₄ in ether gave 3,3-disubstituted azetidine. (**Scheme 12**)

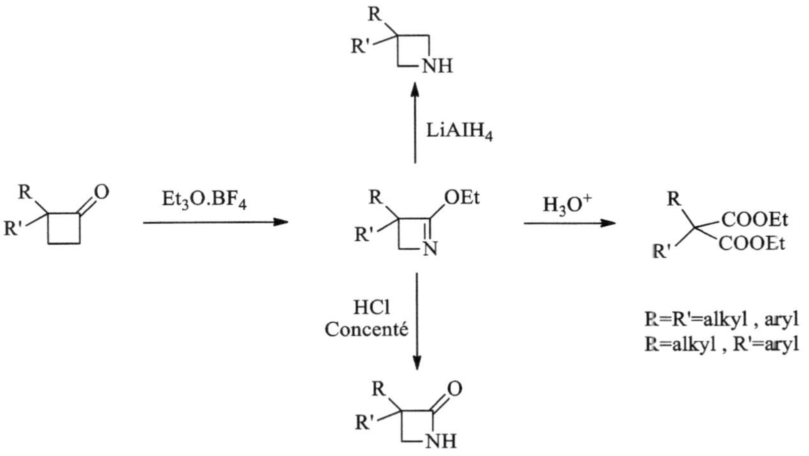

Diagram 12

Thermolysis of 4,4-dimethyl-2-methoxyazetine at 200°C undergoes ring opening to produce methyl acetimidate (E)-N-(prop-1-en-2yl) by 1.5-hydrogen displacement. [5] (**Scheme 13**).

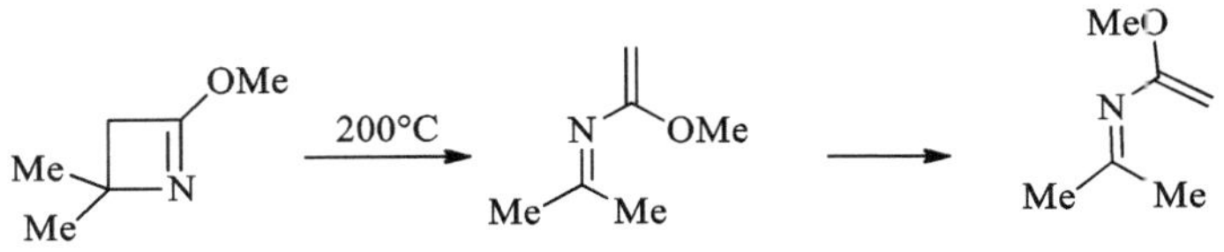

Diagram 13

43

III.3-Five-chain heterocycles :

Pyrrole, furan and thiophene are 5-membered heterocycles containing a single heteroatom. They derive their aromaticity from the delocalization of a single pair of the heteroatom's electron.

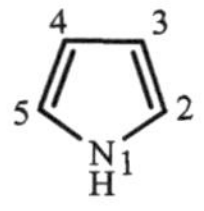

pyrrole 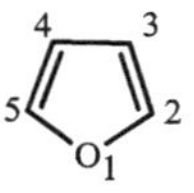   Furane 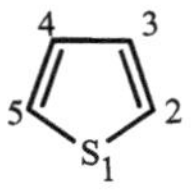   thiophène

For example, furan is the least aromatic of the trio because oxygen has the highest electronegativity and therefore mesomeric represntations that make it relatively less of a contribution to furan's electronic structure than they do in the cases of pyrrole and thiophene. The order of aromaticity is furan $\geq$ pyrrole and thiophene. [11]

Five-membered aromatic heterocycles are considered to be derived from cyclopentadienyl anion 2, and the free doublet of the heteroatom is involved in the delocalization of pi electrons in the ring. The six pi electrons are thus devolved onto five atoms [11] **(Figure 4)**.

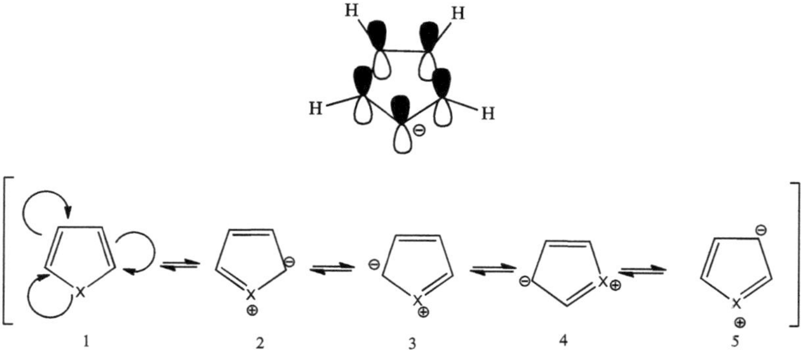

X= NH,S,O

Figure 4: Pi electron delocalization in the 5-membered ring

III.3.1-Pyrrole :

Pyrrole is a five-membered mononitrogenous unsaturated heterocycle first detected by F.F. Runge in 1834 from the distillation of coal tar. It has also been isolated from bone pyrrolysate. It occurs abundantly in nature as a substructure, including chlorophyll, hemoglobin, the dye indigo blue and bile pigments such as bilirubin, biliverdin, chlorine, bacteriochlorin and porphyrin. Pyrrole is also found in many alkaloids as a substructure, produced mainly by plants. Nicotine is one of the best-known pyrrole-containing alkaloids.

III.3.1.1-Physical properties of pyrrole :

Pyrrole is a colorless liquid that darkens easily when exposed to air. It is generally purified by distillation before use. It has a boiling point of 129°C and a specific gravity of 0.967.

Hydrogenation of pyrrole is difficult under acidic conditions, and consequently many reagents used for hydrogenation of aromatic compounds are not applicable to pyrroles.

III.3.1.2 -The chemical properties of pyrrole:

III.3.1.2.1 Aromatic character of pyrrole

Pyrrole has 6 delocalized electrons, counting the free doublet on the nitrogen atom.

(4n+ 2 = 6, n = 1, according to Hückel's rule of thumb). Pyrrole is therefore an aromatic compound. This delocalization of the free nitrogen doublet reduces the electron density around this atom, making the NH group weakly acidic. Resonance occurs between 5 limit forms (**Figure 5**).

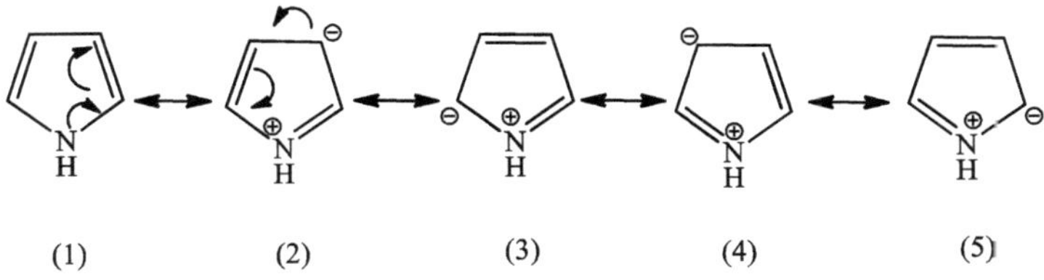

Figure 5: The different mesomeric forms of pyrrole.

According to the rules of mesomerism, the uncharged shape 1 dominates. Similarly, among the charged shapes, those with the smallest charge spacing, boundary shapes 3 and 5, are more representative than boundary shapes 2 and 4, whose charge spacing is greater.

The carbon atoms are partially negatively charged, while the nitrogen atom carries a partial positive charge. The resonance energy of pyrrole is 100 kJ.mol^{-1} . That of benzene is 153 kJ.mol^{-1} . Pyrrole therefore has a lower chemical stability than benzene. [3]

III.3.1.2.2 - Cycle opening :

Very few reactions lead to the opening of the cycle by producing a single compound. Hydroxylamine hydrochloride, in the presence of sodium carbonate, opens the cycle by giving succindialdehyde dioxime with pyrrole. [3] **(Diagram 14)**

Figure 14

III.3.1.2.3 - Actions of electrophilic reagents on Pyrrole :

Electrophilic substitution reactions on cyclic carbons are facilitated by their high electron density. The attack of an electrophilic reagent first leads to intermediates A and B, which then lose a proton to lead to substituted pyrroles.

The intermediate has a more delocalized positive charge (3 boundary forms) than intermediate B (2 boundary forms). The formation of A is therefore favored. **(Diagram 15)**

Figure 15

III.3.1.2.3.1 - Action of strong acids on Pyrrole :

In a strong acid medium (HCI 6N), pyrrole first leads to two cations, C and D

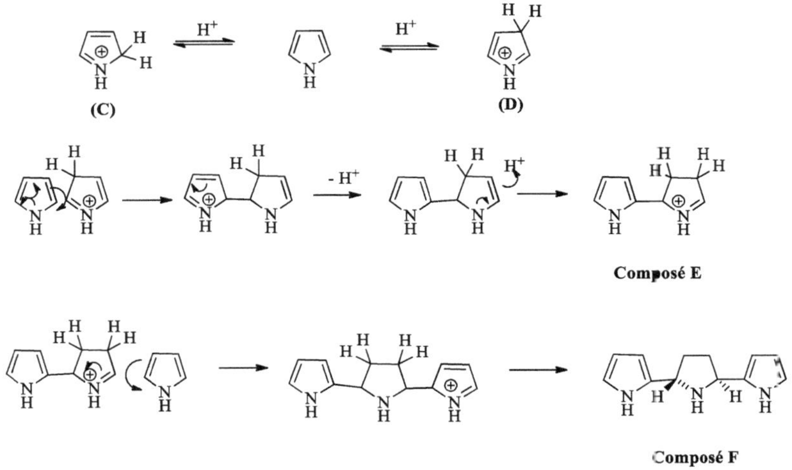

Diagram 16

The majority cation **C**, 2H-pyrrolium ion, thermodynamically the most stable, but much less reactive than cation **D**, 3H-pyrrolium, has a pKa of - 3.6. Cation **D has** a pKa of - 5.9. The latter acts as a powerful electrophilic reagent on an as yet unprotonated pyrrole molecule to give compound **E**, which reacts with another pyrrole molecule to form compound **F**, and so on, finally leading to a "pseudo polymer", a complex of 2 pyrroles and a tetrahydropyrrole **(Scheme 16)**.

III.3.1.2.3.2 -Nitration:

Due to pyrrole's sensitivity to strong acids, nitration is carried out under relatively mild conditions using nitric acid in the presence of acetic anhydride at -10°C . This mixture yields acetylnitrate and acetic acid.

Acetylnitrate leads to a mixture of nitro derivatives in positions 2 and 3. These reactions take place much more easily than with benzene (10^5 times faster). [3] (**Scheme 17)**

$$HNO_3 + Ac_2O \longrightarrow AcONO_2 + AcOH$$

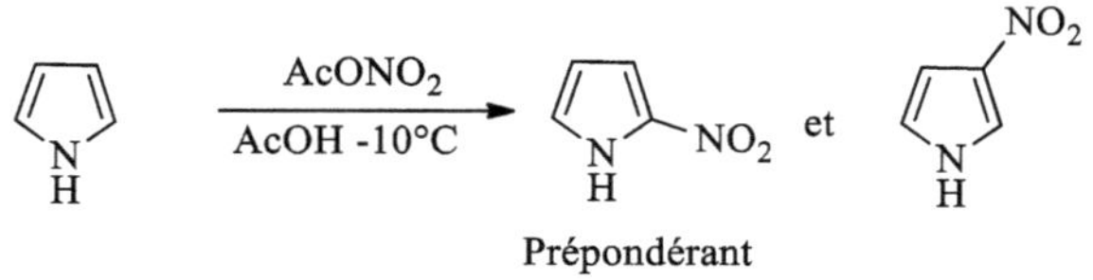

Diagram 17

III.3.1.2.3.3 -Halogenation :

Chlorination is carried out in ether using sulfuryl chloride at 0°C (or sodium hypochlorite). A mixture of derivatives, monochlorous in position 2, and dichlorous in positions 2, 5 is obtained. At higher temperatures, tetrachloro derivatives are formed. Pentachloropyrrolenine may also be the result.

The monobromine derivative in position 2 is prepared on its own by the action of N-bromosuccinimide (NBS). Bromine in acetic acid provides the tetrabromine compounds.

2-monobromo and 2-monochloro derivatives are unstable compounds. N-silyl pyrrole derivatives are brominated by NBS in the 3- and 4-positions. The 3-halopyrroles are stable and behave like aromatic halides. (**Scheme 18**)

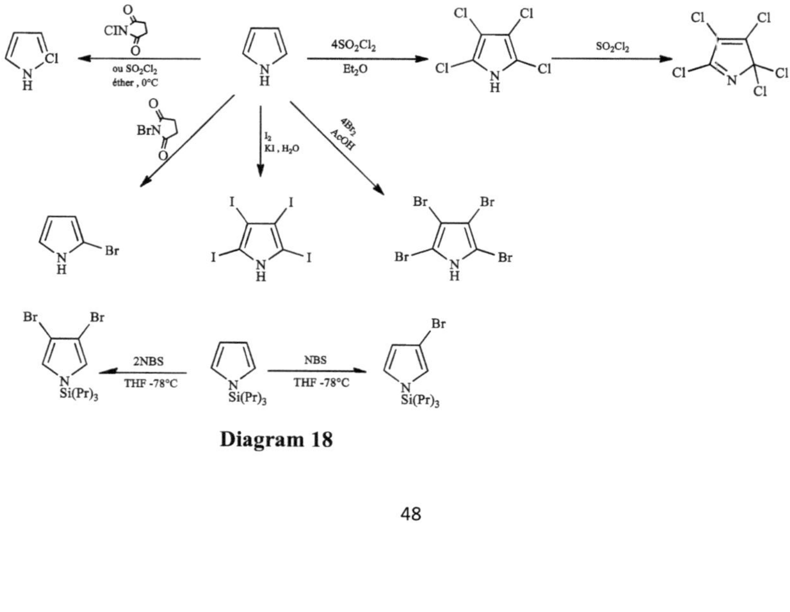

Diagram 18

III.3.1.2.3.4 -Sulfonation :

Pyrrole is highly sensitive to concentrated sulfuric acid and causes polymerization. Sulfonation of pyrrole was achieved using a sulfur trioxide-pyridine complex at 100°C to give pyrrole-2-sulfonic acid **(Scheme 19)**.

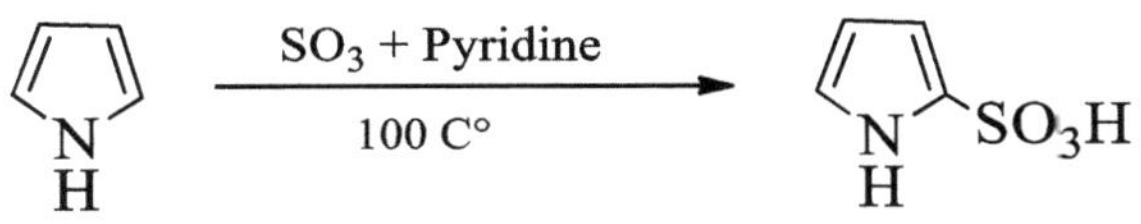

Figure 19

III.3.1.2.3.5 -Reactions with dichlorocarbene :

Depending on the experimental conditions, two mechanisms are possible, corresponding to the formation of singlet or triplet dichlorocarbenes.

In a weakly basic medium, with the generation of dichlorocarbene by heating sodium trichloroacetate in a neutral aprotic solvent, a [2+1] cycloaddition on a double bond of the ring is observed. The bicyclic system obtained by this reaction evolves towards 3-chloropyridine with loss of hydrogen chloride.

In a strongly basic environment, with the generation of dichlorocarbene by the action of potassium hydroxide on chloroform, electrophilic substitution is favored and the result is pyrrole-2-carboxaldehyde. **(Scheme 20)**

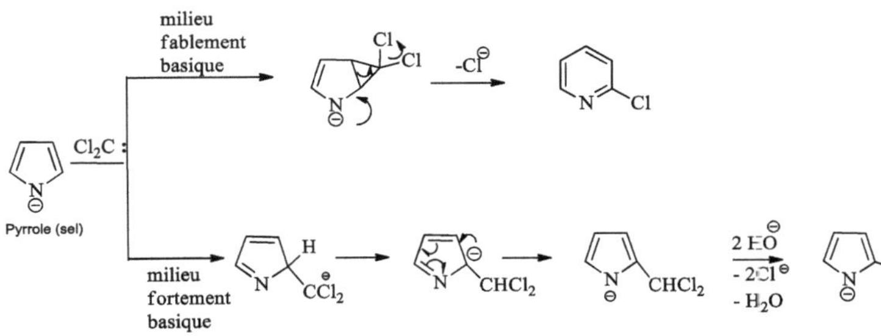

Diagram 20

49

III.3.1.2.3.6 -Acylation :

Pyrrole is directly acylated by heating with acetic anhydride at 200°C *(1)* to give a mixture of 2-acetylpyrrole as the main product.

However, N-acetylpyrrole is obtained in high yields by heating pyrrole with N-acetylimidazole *(2)*.

The reaction of pyrrole with a reactive anhydride such as trifluoroacetic anhydride (TFAA) or trichloroacetic anhydride at room temperature produces the respective acetylpyrroles*(3)* **(Figure 21).**

Diagram 21

III.3.1.2.3.7 -Condensation with aldehydes and ketones :

Condensation of pyrrole with aliphatic aldehydes leads to polymers. Most pyrrole derivatives condense very easily with aliphatic or aromatic ketones and aldehydes in the presence of weak acid.

Hydroxymethylation first forms an unstable pyrrolylcarbinol, which immediately loses a molecule of water in the presence of acid, leading to the 2-alkylidenepyrrolium cation, a powerful electrophile (A), which can be reduced to an alkyl derivative, or reacted with a new pyrrole molecule to form a dipyrrylmethane derivative (B). **(Scheme 22)**

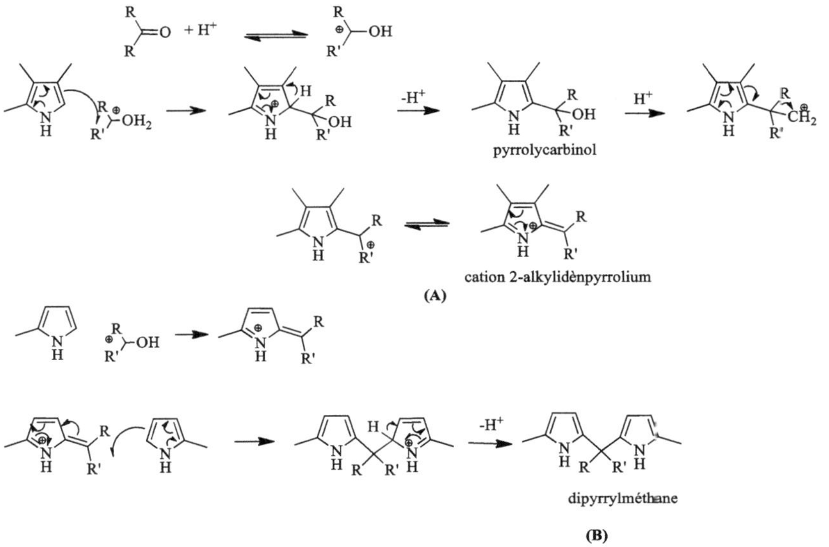
pyrrolycarbinol
cation 2-alkylidènpyrrolium
(A)
dipyrrylméthane
(B)

Diagram 22

III.3.1.2.4 -Nucleophilic substitution reactions:

Due to the nucleophilic nature of pyrrole, nucleophilic substitution is not easy unless electron-withdrawing groups are present in the ring.

1-Methyl-2,5-dinitropyrrole (A) with piperidine (B) at room temperature in dimethyl sulfoxide (DMSO) gave (1-methyl-5-nitro-pyrrol-2-yl) piperidine, while exposure of 2,3-dinitro-1-methylpyrrole (C) with sodium methoxide gave 2-methoxy-3-nitro-1-methylpyrrole in very good yield.

An unprecedented nucleophilic substitution is observed when the parent pyrrole is heated with sodium hydrogen sulfite at 100°C, producing sodium pyrrolidine-2,5-disulfonate presumably via the C-protonated pyrrole[3] . **(Scheme 23)**

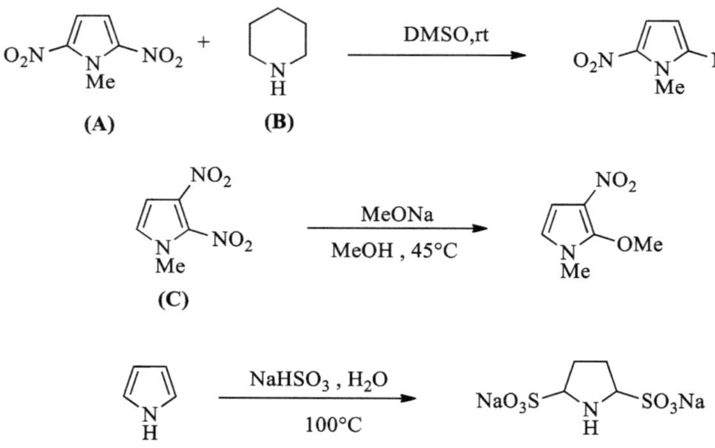

Diagram 23

III.3.1.2.5 -Reduction :

The pyrroles are reduced to pyrrolidines by hydrogen (1 atm, 25°C) in the presence of platinum as catalyst, and at elevated pressure and temperature with Raney nickel (**Scheme 24**).

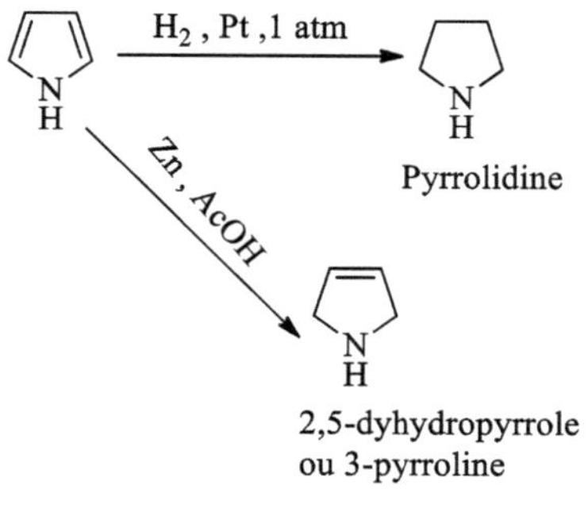

Diagram 24

This reduction is easier if the nitrogen is substituted by an attracting group. The pyrrolic ring is not reduced by sodium in ethanol or ammonia, nor by hydrides. The acidic environment favors hydrogenation. The nascent hydrogen generated by zinc in acetic acid produces 2,5-dihydropyrrole (or 3-pyrroline).

III.3.1.2.6 -Oxidation :

Pyrrole is oxidized by chromium trioxide in acetic acid to maleimide. Hydrogen peroxide, in the presence of barium carbonate at 100°C, oxidizes pyrrole to 3-pyrrolin-2-one (predominant), which is in equilibrium with 2-hydroxypyrrole and 4-pyrrolin-2-one. [5]

(Diagram 25)

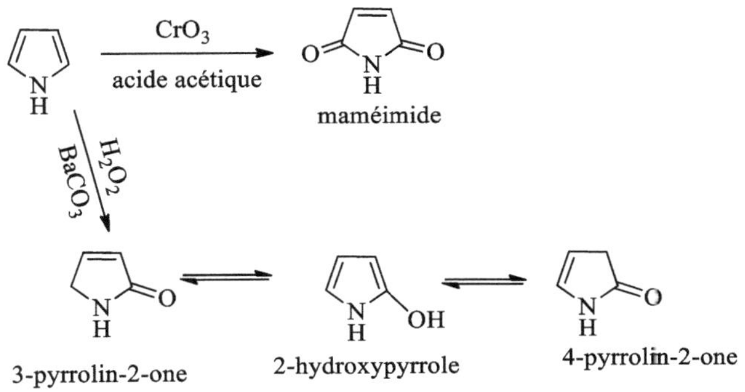

Diagram 25

III.3.1.2.7 -Thermal rearrangement :

Some pyrrole derivatives are transformed at elevated temperatures into positional or ring-extended isomers.

At 600°C, 1-methylpyrrole undergoes thermal rearrangement to produce a mixture of 2-methylpyrrole, 3-methylpyrrole and pyridine. Under similar conditions, 1-phenylpyrrole is also rearranged to a positional isomer, 2-phenylpyrrole **(Scheme 26)**.

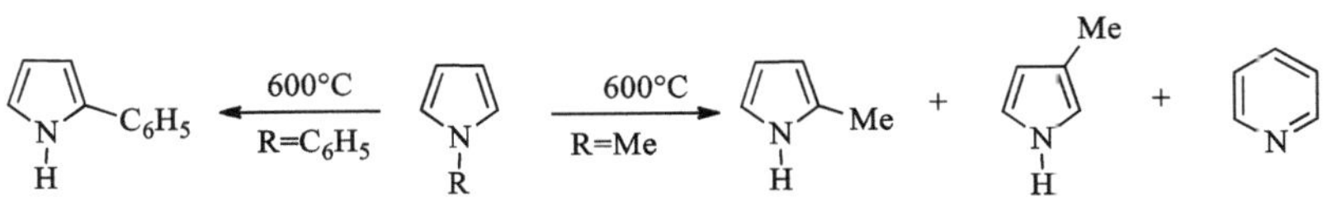

Diagram 26

53

III.4-Six-membered heterocycles :

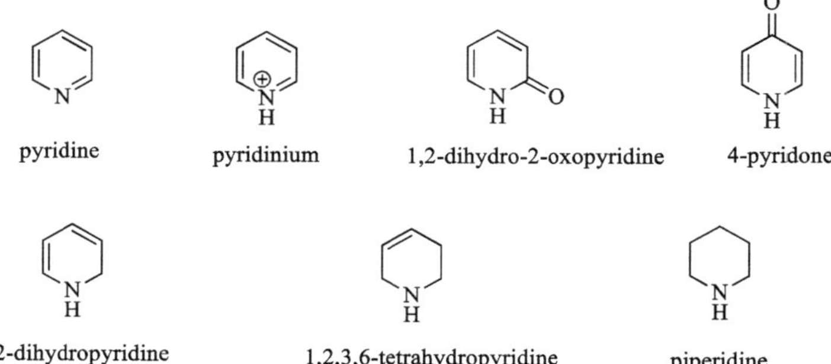

III.4.1 -Aromaticity of six-membered heterocycles :

Since six-membered aromatic heterocycles are considered to be derived from the benzene ring by replacing the CH group with the electronegative heteroatom, the chemical behavior of these heterocycles can be considered similar to that of benzene.

By introducing heteroatoms. The electronegativity of heteroatoms causes electrons to be removed from ring carbon atoms, thus affecting the positioning of the electrons causing a lack of electrons on the circulating carbon atoms. Consequently, six-membered heterocycles are referred to as "n-deficient" heterocycles), and electrophilic substitution reactions are 107 times more difficult than with benzene. [2]

III.4.2 - Aromaticity of pyridine :

We have 3 double bonds and a free doublet, giving a total of 4 pi bonds, or an even number, which means that the molecule is anti-aromatic. let's go back to Hückel's rule, which states that nitrogen participates in aromaticity, but already has a pi bond. This means that these electrons (the free doublet) can't be part of the pi electrons, so they're not counted, so we have an odd number of electron pairs that support our rule that makes pyridine aromatic. [12]

The free doublet of nitrogen, located in a sp-hybridized orbital[2] , is not delocalized, giving this atom a basic character. Pyridine is a weak base, and its low basicity seems at odds with the above observations concerning inductive and mesomeric effects.

The only reason currently given for this low basicity is the hybridization of the nitrogen in the ring. In aliphatic amines or piperidine, which are more basic compounds than pyridine, the nitrogen is hybridized sp^3 and has a lower inductive attracting effect than pyridine nitrogen hybridized sp^2 **(Figure 5)** [3].

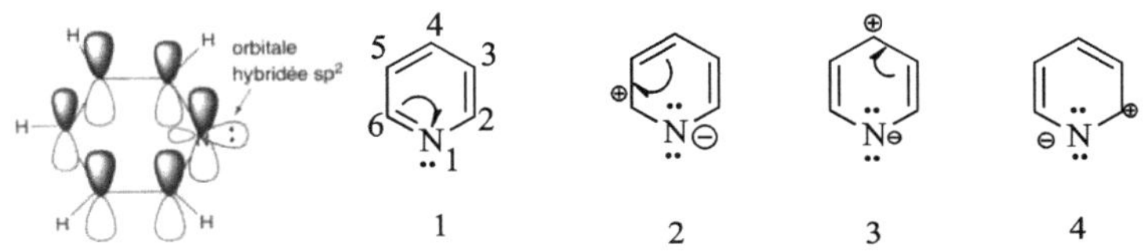

Figure 5: Aromaticity of the pyridine ring

III.4.3-Reactions of electrophilic reagents on cyclic carbons :

The attack of electrophilic reagents on pyridine first affects the nitrogen with the formation of pyridinium ions, making these reactions even more difficult. Pyridine behaves with electrophilic reagents in the same way as nitrobenzene, except for nitration or sulfonation, where the conditions are particularly vigorous and analogous to those used for 1,3-dinitrobenzene (10^{15} times less reactive than benzene).

In the majority of cases, reactions with electrophilic reagents are directed almost exclusively to the 3-position on the pyridine. When an electron-donating group is present in position 3, substitutions by electrophilic reagents are facilitated, and are directed to position 2.

III.4.3.1-Nitration :

This is a very difficult reaction. At 370°C, nitric acid, in the presence of concentrated sulfuric acid, leads to 3-nitropyridine (6%) and 2-nitropyridine (0.5%) (Reaction A).

The presence of inductive donor groups, such as methyl groups, promotes nitration, but also leads to their partial oxidation. Lutidines and collidines are oxidized to the corresponding acids, and decarboxylated to nitropyridines (reactions B and C).

The presence of bulky atoms, such as chlorine, around the nitrogen, in positions 2 and 6, creates a steric hindrance to the formation of pyridinum lions. Combined with the use of nitronium tetrafluoroborate, nitration in this case proves easier. Thus, 2,6-dichloropyridine is nitrated in 77% yield at position 3 (Reaction D). Removal of the chloro groups is carried out by copper in the presence of benzoic acid at 180°C. (**Scheme 27)**

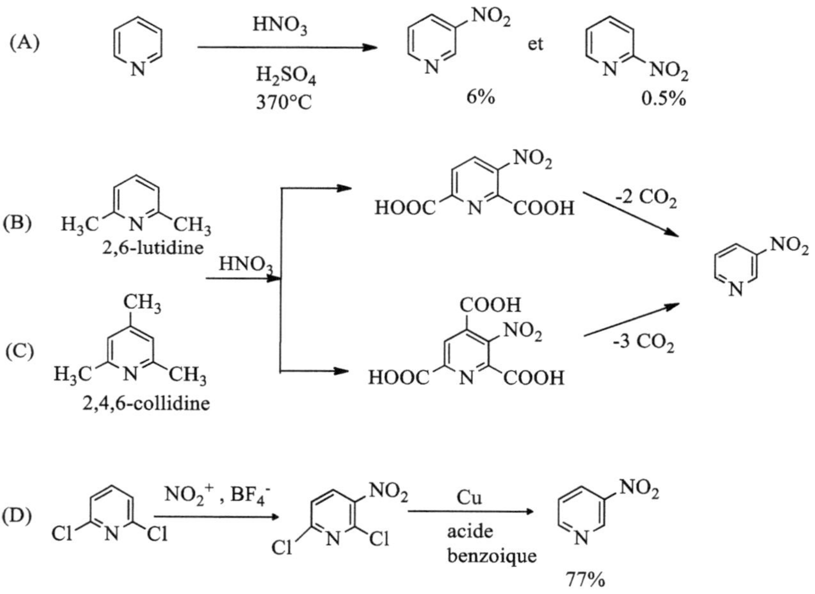

Diagram 27

The presence of a hydroxy or amino group on pyridine facilitates nitration. 2-hydroxypyridine (or 2-pyridone) is nitrated in position 4 (Reaction E).

4-aminopyridine is first nitrated on the amine function to N-nitroamine, which is then rearranged to 4-amino-3-nitropyridine (Reaction F). The reaction is carried out using a mixture of nitric and sulfuric acid at 70°C **(Scheme 28)**.

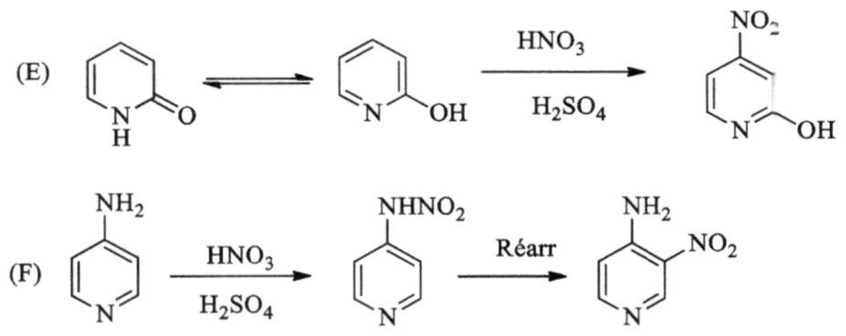

Figure 28

III.4.3.2-Sulfonation :

Pyridine is a polar liquid miscible with organic solvents and water. It can be formally derived from benzene by replacing the CH group with a nitrogen atom.

Pyridine is a highly aromatic heterocyclic ring, but the role of heteroatoms makes its chemical properties completely different from benzene. If there are 6 pi electrons, the heterocyclic ring will be considered aromatic, but these electrons are already intact, excluding the free nitrogen doublet, so the free doublet will be connected without destroying the aromaticity of the ring.

It is considered the alkaline form (pka = 5.2). The pyridine salt can be formed from pyridine and alkyl 1. Pyridine forms a complex by adding Lewis acid such as sulfur trioxide 2 **(Scheme 29)** [11].

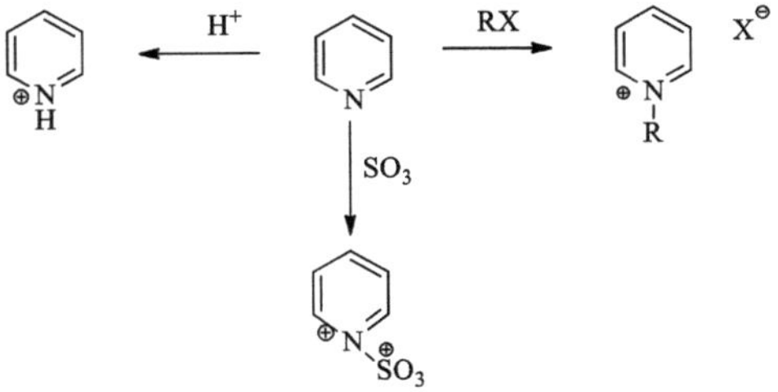

Diagram 29

III.4.3.3-Halogenation :

Bromine in the oleum reacts with pyridine to give 3-bromopyridine (A) in over 80% yield. First, pyridinium-1-sulfonate is formed, which then reacts with bromine.

Bromine or chlorine react between 200 and 300°C (100°C in the presence of aluminum chloride) to give 3-bromo or 3-chloropyridine and 3,5-dibromo or 3,5-dic hloropyridine (B).

At higher temperatures, substitution takes place in position 2, then in positions 2 and 6.

In the presence of the pyridine-palladium hydrochloride complex, chlorine and bromine react at - 5°C to give 2-bromo and 2-chloropyridines (C). **(Scheme 30)**

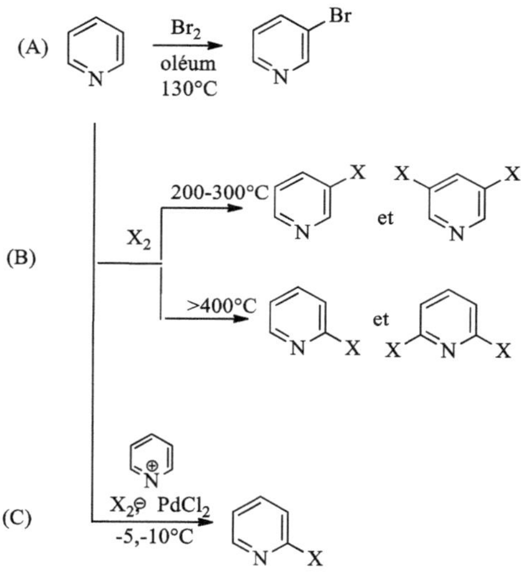

Diagram 30

III.4.3.4-Mercuration :

Pyridine forms a salt when treated with an aqueous solution of mercuric acetate. It is transformed at 180°C into 3-acetoxymércuripyridine. In the presence of sodium chloride, chloromercuric drift is obtained. **(Scheme 31)**

Diagram 31

III.4.4-Reactions with nucleophilic reagents :

Unlike reactions with electrophilic reagents, which are difficult or even impossible with pyridine, those with nucleophilic reagents are numerous and facilitated by the presence of the ring's azomethine bond, which is electron-withdrawing, directing the attack to position 2, or 6, then to position 4 (or vice versa in a few cases), and very rarely to position 3. This attack is followed by the loss of a hydride ion, which usually requires the presence of an oxidizing agent acting as acceptor.

If a leaving group such as the chloro group is present in position 2 or 4, the nucleophile's attack is instead directed towards these carbons with removal of this group, as the hydride ion is a very poor leaving group that requires more vigorous experimental conditions to be removed. **(Scheme 32)**

nécléophiles: RLi, AlH_4^-, NH_2^-, HO^-, RS^-, RO^-, amines et NH_3

Diagram 32

III.4.4.1-Alkylation and arylation :

Addition of alkyl or aryllithium to pyridine leads to lithium salts of dihydropyridines (A), which can sometimes be isolated. This is the case with phenyllithium (reaction carried out at 0°C.

Then, the loss of a lithium hydride molecule through heating is possible, resulting in substitute pyridines. The latter are also formed from 1,2-dihydropyridines obtained by adding dilute acid to lithium derivatives (B).

Air oxidation transforms them into pyridines. Pyridines are more easily alkylated by organolithiums than by organomagnesiums. **(Scheme 33)**

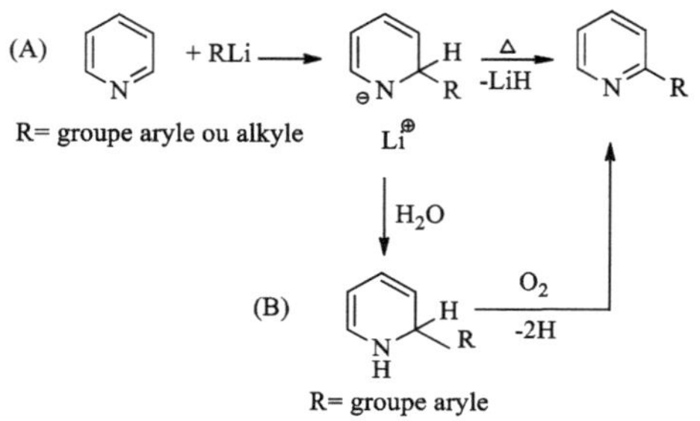

(A)   + RLi ⟶ $\overset{\Delta}{\underset{-LiH}{\longrightarrow}}$

R= groupe aryle ou alkyle

H_2O

(B) $\overset{O_2}{\underset{-2H}{\longrightarrow}}$

R= groupe aryle

Diagram 33

III.4.4.2-Amination :

The action of sodium, potassium or barium amide on pyridine leads to 2-aminopyridine. The reaction can be carried out dry, but more often in boiling aromatic solvents such as toluene or N,N-dimethylaniline, at temperatures above 100°C (heterogeneous conditions).

It can also take place at low temperature in the presence of sodium amide, under homogeneous conditions, in a suitable solvent (optimal conditions). The reaction begins by attacking the amidide anion with nitrogen to form a sodium salt of 2-amino-1,2-dihydropyridine. This salt eliminates a hydrogen molecule before the addition of water liberates 2-aminopyridine. Aromatization is facilitated by the presence of oxidizing agents such as potassium permanganate. This mechanism is the subject of much controversy, and variants exist. The only certainty is the formation of the 2-amino-1,2-dihydropyridine salt and the evolution of hydrogen **(Scheme 34).**

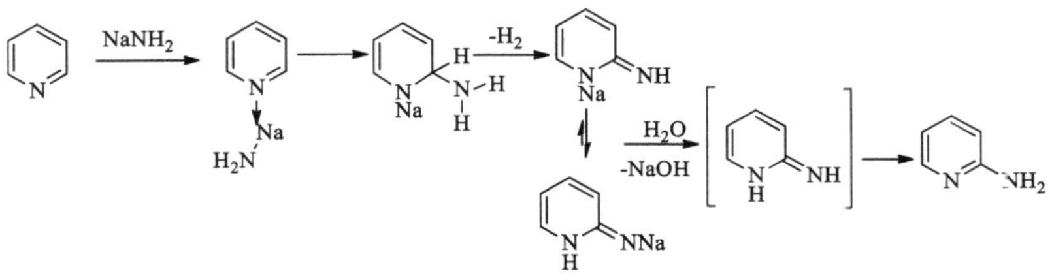

Diagram 34

III.4.4.3-Hydroxylation :

Reactions of potash or sodium hydroxide with pyridine (A) take place under more vigorous conditions than with sodium amidide, as the OH ion⁻ is a weaker nucleophilic reagent than the amidide ion NH_2^- .

After acidification of the anion formed, and oxidation of the resulting 2-hydroxy-1/2-dihydropyridine, 2-pyridone is obtained in low yield. **(Scheme 35)**

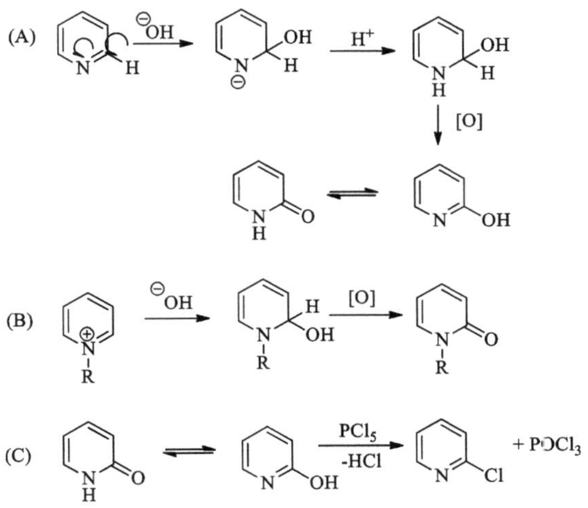

Figure 35

These reactions occur much more easily with pyridinium salts (B). Under the action of phosphorus pentachloride, 2-pyridone leads to OH substitution⁻ of the tautomeric form by Cl, giving 2-chloropyridine (C). [3]

III.5-Condensed aromatic heterocycles :

 The fusion of a benzene to an aromatic heterocycle retains the aromaticity in modified form. These aromatic heterocycles show alternating bonds, suggesting partial positioning of the bonds.

III.5.1-Fused 5-membered aromatic heterocycles :

 From 1 to 4, alternating free doublets can be detected. But in 5-7 where a benzene ring is fused to a five-membered aromatic heterocycle by a single carbon-carbon bond are less aromatic due to the considerable alternation of double bonds **(Figure 6).**

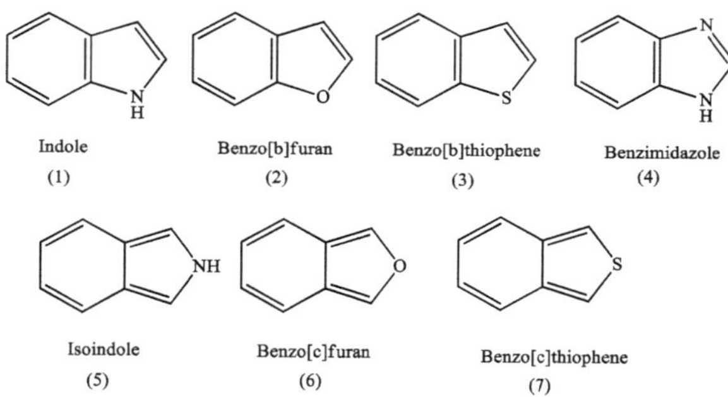

Figure 6: Condensed 5-membered aromatic heterocycles

III.5.2-Condensed aromatic 6-membered heterocycles :

 These heterocycles relate to naphthalene in the same way that pyridine relates to benzene. Fusion of a benzene ring, however, results in a decrease in aromaticity due to alternating bonds **(Figure 7).**

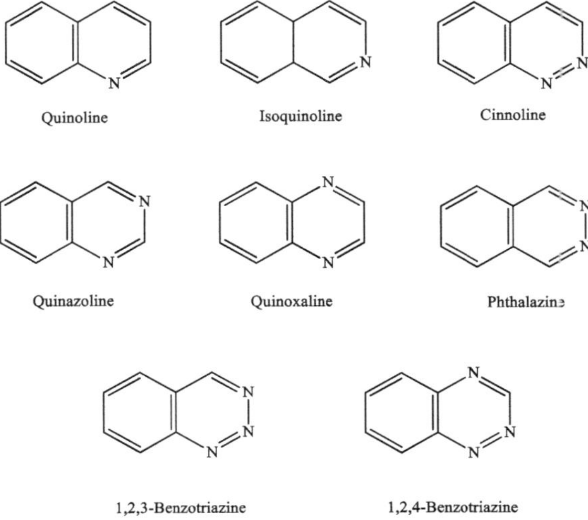

Figure 7: 6-membered condensed aromatic heterocycles

III.5.3-Other condensed heterocycles :

Condensation of two six-membered aromatic heterocycles by a single carbonate bond leads to the formation of heterocyclic naphthalene analogues (A) and (B).

Similarly, the condensation of a six-membered aromatic heterocycle with a five-membered aromatic heterocycle forms a bicyclic heterocycle (C), which is also considered part of the aromatic heterocycle class. [2] **(Figure 8)**

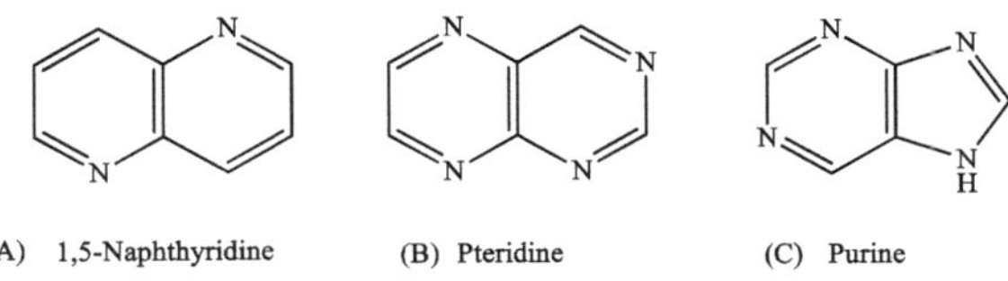

Figure 8

IV-Conclusion :

In this chapter we have studied the different properties of each class of 3- to 6-membered heterocycles; physical priorities are important as criteria for judging heterocycle purity. Our study is based mainly on five- and six-membered heterocycles, since they have satisfactory chemical reactivity and interesting biological and therapeutic physicochemical activities.

CHAPTER III

USE OF HETEROCYCLES FOR THERAPEUTIC PURPOSES

I-Introduction :

Most pharmaceutical products are based on heterocycles, and among the various clinical applications, heterocyclic compounds have a considerable active role as antibacterial, antiviral, antifungal, anti-inflammatory and anti-tumor drugs.

Heterocycles are a structural unit common to most marketed drugs. Of the top five retail sales of small-molecule drugs in the US in 2014, four are or contain heterocycle fragments in their overall structure **(Figure 1)**[13]

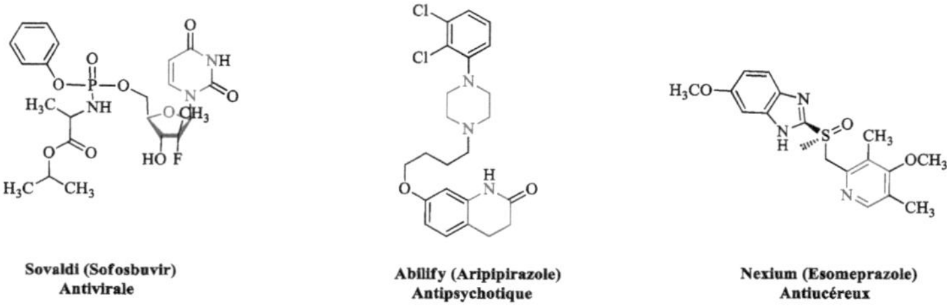

Figure 1: top retail sales of heterocycle small molecule drugs in the US in 2014.

II-Nitrogen heterocycles :

Figure 2: Some examples of drugs containing a nitrogen heterocycle as a main nucleus

II.1-Pyridine :

Many drugs and pesticides contain a pyridine moiety. Examples include antimicrobial agents, antiviral agents, antioxidants, antidiabetic agents, antimalarial agents, anti-inflammatory agents, psychopharmacological antagonists and antiamoebic agents. [13]

These pyridine fractions play an essential role in medicinal chemistry. One of pyridine's roles in medicinal chemistry is to improve water solubility due to its low basicity.

II.1.1-Sulfapyridine :

Sulfapyridine has good antibacterial activity and solubility in water under acidic conditions.

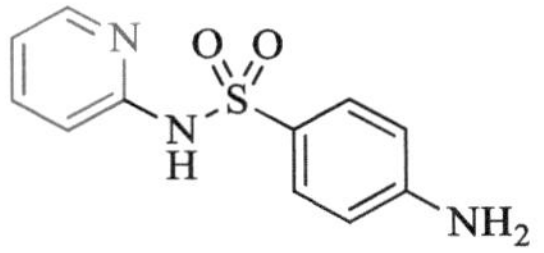

Sulfupyridine

II.1.2-Omeprazole :

Omeprazole is used to treat certain stomach and esophageal problems (such as acid reflux and ulcers). It works by reducing the amount of gastric acid produced. It relieves symptoms such as heartburn, difficulty swallowing and persistent coughing. Helps heal acid damage to the stomach and esophagus. Helps prevent ulcers and may help prevent esophageal cancer. [14]

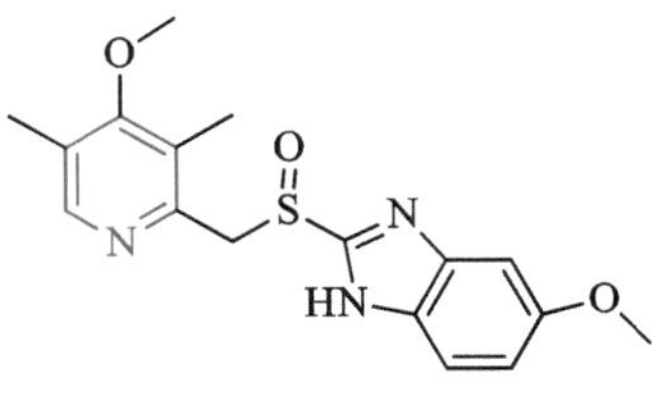

Omeprazole

II.1.3 -Eszopiclone is used as a hypnotic and sedative in the treatment of insomnia

Eszopiclone

II.2-Indole :

Indoles are found in abundance in biologically active compounds such as pharmaceuticals, agrochemicals and alkaloids. The importance of the indole has led to the development of various bioactive compounds by varying the substituents at different positions on the indole ring. Such compounds have been reported for various biological activities such as antimicrobial, antiviral, insecticidal, analgesic, etc. [15]

II.2.1- 2,3- diaylindole :

The indole scaffold is one of the most prevalent heterocycles in natural and synthetic bioactive compounds, including anticancer agents. Due to its biodiversity and versatility, it has been a highly favored motif for the targeted design and development of anticancer agents-let's talk about 2,3- diaylindole.

2,3-diarylindole

II.2.2-Indometacin :

Indomethacin, also known as indomethacin, is a non-steroidal anti-inflammatory drug commonly used as a prescription medication to reduce fever, pain, stiffness and swelling caused by inflammation.

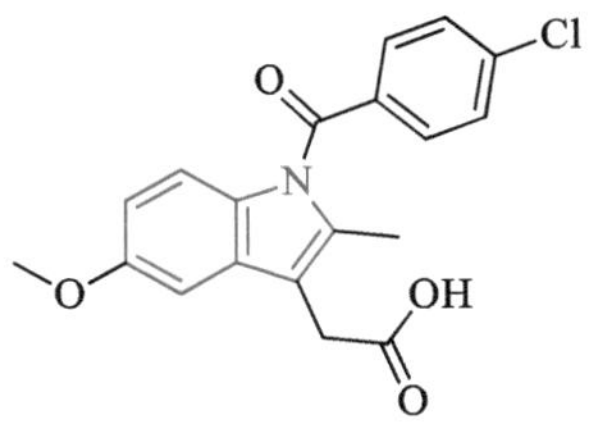

Indometacine

II.3-Pyrimidine:

Some pyrimidine derivatives are primary structural units of DNA and RNA, underscoring the fact that pyrimidine belongs to a special class of compounds for drug research and development [16].

II.3.1-Methotrexate :

Many anti-cancer drugs contain the pyrimidine cycle. One of the first drugs, still used today, is methotrexate, which acts by inhibiting the formation of folic acid. [6]

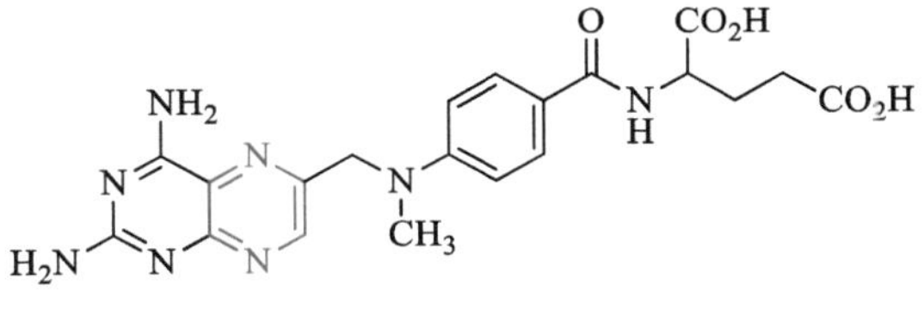

Methotrexane

II.3.2- Risperidone :

Risperidone is an atypical antipsychotic widely used in the treatment of mania and schizophrenia. Risperidone treatment is associated with elevations in serum aminotransferases and, in rare cases, has been linked to clinically apparent acute liver injury. [17]

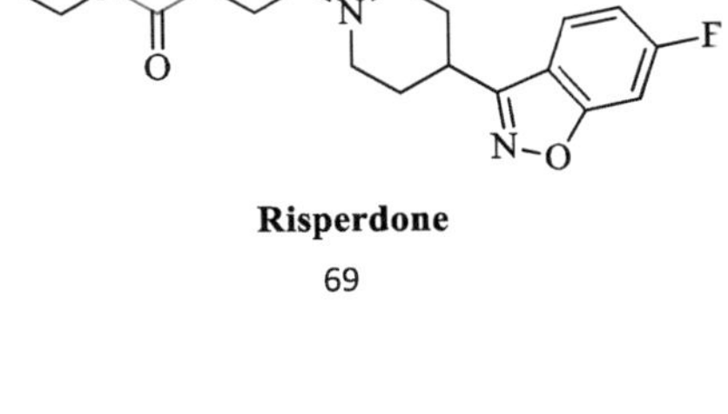

Risperdone

III-Oxygenated heterocycles :

Oxygen heterocycles are the second most common type of heterocycles that appear as structural components of U.S. Food and Drug Administration (FDA)-approved pharmaceuticals. Analysis of our database of approved drugs through 2017 reveals 311 distinct pharmaceutical products containing at least one oxygen heterocycle. [18] **Figure 3**

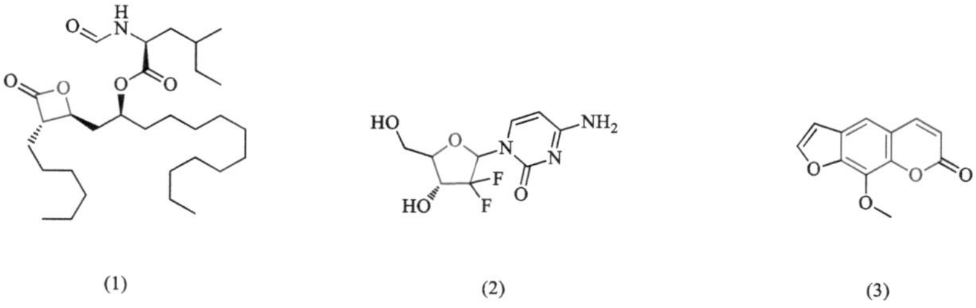

(1)   (2)   (3)

Figure 3: Some examples of drugs containing an oxygenated heterocycle as a main nucleus

III.1-Oxirane :

The oxirane (epoxide) functional group is arguably the most synthetically useful small-ring heterocycle due to its ease of synthesis and large-scale ring-opening reactions, which generally occur with predictable regioselectivity and stereospecificity. [19]

III.1.1-Scopolamine :

Scopolamine, is an anticholinergic drug used to treat a variety of conditions such as nausea, peptic ulcers, irritable bowel syndrome and chronic obstructive pulmonary disease (COPD). [20]

Scopolamine

III.1.2-Natamycin :

Natamycin is an antifungal available in eye drops to treat fungal infections around the eye[17].

Natamycin

III.2-Furan :

III.2.1-Dantrolene :

This drug is used to treat muscle stiffness and cramps (spasms) caused by certain nerve disorders such as spinal cord injury. [21]

Dantrolene

III.2.2-Furosemide :

Furosemide is administered to help treat fluid retention (edema) and swelling caused by congestive heart failure, liver disease, kidney disease or other medical conditions.

Furosemide

III.2.3-Furadantine :

Furadantine is an antibacterial agent used to treat urinary tract infections[17].

Furadantine

III.3- Benzofuran:

II.3.1-Methoxsalen :

Methoxsalen is used as a treatment called PUVA to treat vitiligo, a condition in which skin color is lost, and psoriasis, a skin condition associated with red, scaly patches[17].

Methoxsalen

III.3.2-Amiodarone :

Amiodarone is an antiarrhythmic drug used to treat ventricular tachycardia or ventricular fibrillation[17].

Amiodarone

IV-Sulfur heterocycles :

Heterocycles containing sulfur as a heteroatom are one of the most important classes of heterocyclic compounds in continuous use in chemistry, with numerous pharmacological actions such as anticancer, antiviral, anti-inflammatory, antimicrobial, antituberculosis, [22] etc...**Figure** 4

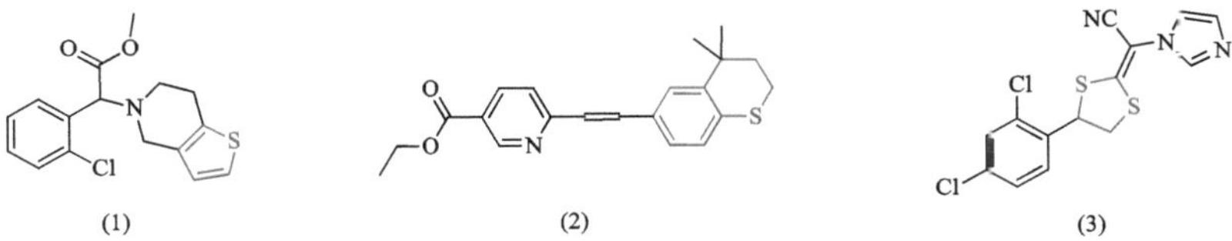

(1)   (2)   (3)

Figure 4: Some examples of drugs containing a sulfur-containing heterocycle as a main nucleus

IV.1-Thiophene :

IV.1.1-Suprofen :

Suprofen acts as a non-steroidal anti-inflammatory, non-narcotic analgesic, anti-rheumatic, peripheral nervous system drug and drug allergen.

Suprofene

IV.1.2-Articaine :

Articaine is a dental amide-type local anesthetic. It is the most widely used local anesthetic in a number of European countries, and is available in many others. It is the only local anesthetic to contain a thiophene ring.

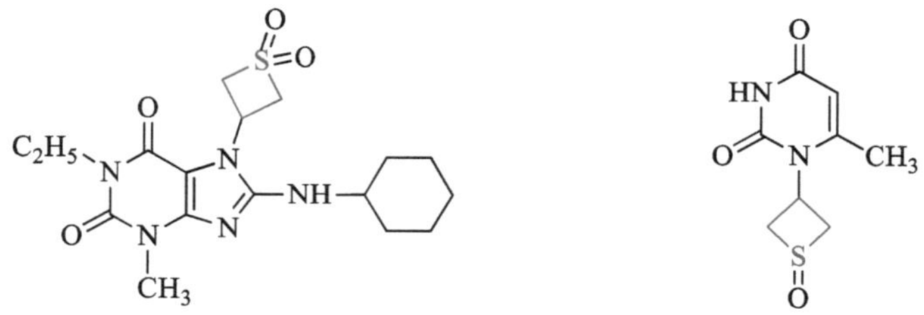

Articaïne

IV.2- Thetane :

Thetane is found in many drugs, both as antidepressants and antihypertensives.

Activité antidepréssive   **Activité antihypertenssive**

IV.3- Thiopyran :

In 2019, a new series of Thiopyran derivatives containing a bicyclopyrazolone moiety was reported, then evaluated for antibacterial and antifungal activity.

- *Antifungal activity :*

- *Antibacterial activity :*

V-Conclusion :

In this chapter, we have set out the biological activities of heterocycles, which have been of great use in industrial pharmaceutical chemistry and have attracted our attention, elucidating the role of each family in certain drugs.

We've found that the biological activity of these heterocycles is closely linked to the number of chain members and the heteroatom present, so there's multiple activity for each type of heterocycle.

Various studies have shown that heterocycles condensed with benzene exhibit more interesting biological activity.

GENERAL CONCLUSION

GENERAL CONCLUSION

Heterocyclic reactions have understandably attracted the attention of many research groups, both in the academic field and in the pharmaceutical industry.

We can now synthesize them using new types of heterocyclic structures and methods, and there's no doubt that heterocyclic chemistry will continue to progress. And science has taught us how to use heterocycles to improve the quality of human life and explore the secrets of nature.

Heterocycles are chemically more flexible and structurally more rigid to meet the many requirements of biochemical systems.

The speed at which heterocyclic compounds continue to be invented testifies to the strength and vitality of this field of organic chemistry. The challenges of discovering new heterocyclic systems and understanding their properties also continue to drive research in the field.

Bibliography

[1] AOUMEUR.N ; *"Mémoire de magister"*. University of Oran 1, February **2011**.

[2] Gupta, R. R., Kumar, M., & Gupta, V *Heterocyclic Chemistry: Volume II: Five-Membered Heterocycles*. Springer Science & Business Media. . **(2013)**.

[3] R. MILCENT, Heterocyclic Organic Chemistry, EDP Sciences, **2003.**

[4] G. Bélanger. Heterocyclic organic chemistry course 706, September **2018**
[5] Ram, Vishnu Ji, et al. *The Chemistry of Heterocycles: Nomenclature and Chemistry of Three to Five Membered Heterocycles*. Elsevier, **2019**.

[6] Tyrell, John A., and Louis D. Quin. *Fundamentals of heterocyclic chemistry: importance in nature and in the synthesis of pharmaceuticals*. John Wiley & Sons, **2010.**

[7] Eicher, Theophil, S. Hauptmann, A.Speicher. *The chemistry of heterocycles: structures, reactions, synthesis, and applications*. John Wiley & Sons, **2013**.

[8] *"Saturated Heterocyclic Compounds." Saturated Heterocyclic Compounds Biotechnology and Biomaterials*, https://www.omicsonline.org/saturated-heterocyclic-compounds-peer-reviewed-open-access-journals.php.

[9] Aromaticity. Wikipedia, the free encyclopedia. Page accessed 07:47, March 21, **2019** from http://fr.wikipedia.org/w/index.php?title=Aromaticit%C3%A9&oldid=157734418.

[10] "Huckel Rule." June 5, **2019.** https://chem.libretexts.org/@go/page/126935

[11] Jones, R. C. F. "Aromatic heterocyclic chemistry: David T. Davies. Davies, Oxford Science Publications, **1992**

[12] "Aromaticity Rules + Cyclic, Charged & Heterocyclic Aromatic Compounds." MCAT and Organic Chemistry Study Guides & Tutoring, Feb. 18, **2016**, https://leah4sci.com/aromaticity-tutorial-for-cyclic-charged-and-heterocyclic-aromatic-compounds

[13] Martins, Pedro et al. *"Heterocyclic Anticancer Compounds: Recent Advances and the Paradigm Shift toward the Use of Nanomedicine's Tool Box." Molecules (Basel, Switzerland)* . 16 Sep. **2015**, doi: 10.3390/molecules200916852

[14] Omeprazole Oral: Uses, Side Effects, Interactions, Pictures, Warnings & Dosing - WebMD.   https://www.webmd.com/drugs/2/drug-3766-2250/omeprazole-oral/omeprazole-delayed-release-tablet-oral/details

[15] Sravanthi, T.V., Manju, S.L., Indoles - *A promising scaffold for drug development*, **(2016)**, doi: 10.1016/j.ejps.2016.05.02

[16] Dinesh, Reddy & Kongot, Manasa & Kumar, Amit **(2021)**. *Coumarin hybrid derivatives as promising leads to treat tuberculosis: Recent developments and critical aspects of structural design to exhibit anti-tubercular activity.*

[17] National Center for Biotechnology information **(2022)**.

[18] J. Med. Chem - DOI: 10.1021/acs.jmedchem.8b00876 - Publication Date (Web): 19 Jul 2018

[19] E. Gras, O. Sadek, - *Oxiranes and Oxirenes: Fused-Ring Derivatives,*

Publisher(s): D. Black, J.Cossy, C. Stevens, *Comprehensive Heterocyclic Chemistry IV*, Elsevier, *https://doi.org/10.1016/B978-0-12-818655-8.00026-3*

[20] A. Altaf, A.Shahzad, Z.Gul, N.Rasool, A. Badshah, B. Lal, E. Khan. *A Review on the Medicinal Importance of Pyridine Derivatives. Journal of Drug Design and Medicinal Chemistry*. Vol. 1, No. 1, **2015**. doi: 10.11648/j.jddmc.20150101.11

[21] Pozharskii, Alexander F., Anatoly Timofeevich Soldatenkov, and A. R. Katrizky. *"Heterocycles in Life and Society, an Introduction to Heterocyclic Chemistry and Biochemistry and the Role of Heterocycles in Science, Technology, Medicine and Agriculture." European Journal of Medicinal Chemistry (***1997***).*

[22] Pinder, R. M., Brogden, R. N., Speight, T. M., & Avery, G. S Dantrolene sodium. *Drugs*, (**1977**).

[23] Pathania S, Narang RK, Rawal RK. *Role of Sulphur-heterocycles in medicinal chemistry: An update*. Med Chem. **2019** doi: 10.1016/j.ejmech.2019.07.043.

Printed by Books on Demand GmbH, Norderstedt / Germany